T0145327

Embedded Deep Learning

Bert Moons • Daniel Bankman • Marian Verhelst

Embedded Deep Learning

Algorithms, Architectures and Circuits
for Always-on Neural Network Processing

 Springer

Bert Moons
ESAT-MICAS
KU Leuven
Leuven, Belgium

Daniel Bankman
Department of Electrical Engineering
Stanford University
Stanford, CA, USA

Marian Verhelst
ESAT-MICAS
KU Leuven
Leuven, Belgium

ISBN 978-3-030-07577-4   ISBN 978-3-319-99223-5   (eBook)
https://doi.org/10.1007/978-3-319-99223-5

This Springer imprint is published by the registered company Springer Nature Switzerland AG
The registered company address is: Gewerbestrasse 11, 6330 Cham, Switzerland

To Lise

Preface

Although state of the art in many typical machine learning tasks, deep learning algorithms are very costly in terms of energy consumption, due to their large amount of required computations and huge model sizes. Because of this, deep learning applications on battery-constrained wearables have only been possible through wireless connections with a resourceful *cloud*. This setup has several drawbacks. First, there are privacy concerns. Cloud computing requires users to share their raw data—images, video, locations, speech—with a remote system. Most users are not willing to do this. Second, the cloud-setup requires users to be connected all the time, which is unfeasible given current cellular coverage. Furthermore, real-time applications require low latency connections, which cannot be guaranteed using the current communication infrastructure. Finally, wireless connections are very inefficient—requiring too much energy per transferred bit for real-time data transfer on energy-constrained platforms. All these issues—privacy, latency/connectivity, and costly wireless connections—can be resolved by moving towards computing at the *edge*. Finding ways to do this is the main topic of this dissertation. It focuses on techniques to minimize the energy consumption of deep learning algorithms for embedded applications on battery-constrained wearable edge devices.

Computing in the edge is only possible if these deep learning algorithms can be run in a more energy-efficient way, within the energy and power budget of the computing platform available on a wearable device. In order to achieve this, several innovations are necessary on all levels of an application's design hierarchy. Smarter **applications** can be developed for more statistically efficient deep learning **algorithms**, which in turn should run on optimized **hardware** platforms built on specifically tailored **circuits**. Finally, designers should not focus on any of these fields separately but should **co-optimize** hardware and software to create minimum energy deep learning platforms. This book is an overview of possible solutions towards designing such systems.

Leuven, Belgium
Stanford, CA, USA
Leuven, Belgium
May 2018

Bert Moons
Daniel Bankman
Marian Verhelst

Acknowledgments

The authors would like to thank several people and institutions for their valuable contributions to this work.

We thank Tom Michiels, Edith Beigne, Boris Murmann, Wim Dehaene, Tinne Tuytelaars, Ludo Froyen, and Hugo Hens for their comments and discussion. IWT, intel, Qualcomm, and Synopsys for their funding, software, and support. Thanks to Florian Darve and Marie-Sophie Redon at CEA-LETI and to Etienne Wouters and Luc Folens at imec/IC-Link for their back-end support in ASIC design.

We would also like to acknowledge all coauthors for their contributions to this manuscript. Thanks to Koen Goetschalckx, Nick Van Berckelaer, Bert De Brabandere, Lita Yang, Roel Uytterhoeven, Martin Andraud, and Steven Lauwereins.

Contents

Acronyms

as	Accuracy-scalable
AC	Approximate computing
ANN	Artificial neural network
ANT	Algorithmic noise tolerance
ASIC	Application-specific integrated circuit
ASIP	Application-specific instruction-set processor
BW	Bandwidth
CDAC	Capacitive digital to analog converter
CM	Common-mode
CNN	Convolutional neural network
CONVL	Convolutional layer
CPU	Central processing unit
CSA	Carry save adder
DAS	Dynamic-accuracy-scaling
DVAS	Dynamic-voltage-accuracy-scaling
DVAFS	Dynamic-voltage-accuracy-frequency-scaling
EDA	Electronic design automation
EDP	Energy delay product
FCL	Fully connected layer
FCN	Fully connected network
FPNN	Fixed-point neural networks
GB	Gigabyte
GOPS	Giga-operations per second
GOPS/W	Giga-operations per second per Watt
GPU	Graphical processing unit
HPC	High-performance computing
I2l	Input-to-label
IC	Integrated circuit
IoT	Internet of Things
ISA	Instruction set architecture
LSB	Least significant bit

LSTM	Long short-term memory
MAC	Multiply-accumulate
MSB	Most significant bit
MSBNN	Mixed-signal binary neural network ASIC
NLP	Natural language processing
NPU	Neural processing unit
nas	Non-accuracy-scalable
nvas	Non-voltage-accuracy-scalable
QNN	Quantized neural networks
RMS	Recognition, mining, and synthesis
RNN	Recurrent neural networks
SC	Switched-capacitor or switch-cap
SGD	Stochastic gradient descent
SotA	State of the art
STE	Straight-through estimator
TOPS	Tera-operations per second
TOPS/W	Tera-operations per second per Watt
V_t	Threshold voltage of CMOS transistors
VDD	Typical name for the supply voltage
VLIW	Variable length instruction word
VOS	Voltage over-scaling
vas	Voltage-accuracy-scalable

Chapter 1
Embedded Deep Neural Networks

1.1 Introduction

Humankind has long dreamed of creating machines that think. Ever since the conception of programmable computers, people have wondered whether these machines could become intelligent. Today, some of these goals in creating **artificial intelligence** have been achieved. Intelligent software can automate intuitive tasks such as understanding and translating speech (Chiu et al. 2017) and interpreting images (Krizhevsky et al. 2012a), even in order to make reliable image-based diagnoses in medicine (Esteva et al. 2017). Solutions to less intuitive problems in AI, such as human-level machine intelligence and consciousness, are still very far off.

The class of more intuitive problems, such as speech and image recognition, is currently solved through letting computers learn from experience and by modeling reality through a hierarchy of concepts. The mathematical representation of these models typically is a deep graph, with many layers. Hence, for this reason, this set of state-of-the-art (SotA) techniques is generally referred to as deep learning. More specifically, deep learning uses neural networks with many layers to represent these abstract models. Neural networks have been a source of study since 1940s, but have only recently broken through and advanced the SotA in a number of recognition tasks. Since 2012, deep learning algorithms have shown unprecedented performance in a number of tasks and have broken record after record in AI challenges and competitions. As of 2014, they have surpassed the human performance in visual recognition and since 2016 in speech recognition, as shown in Fig. 1.1. Now, for the first time in human history, we have developed machines that can reliably replace some of our senses and the cognitive ability necessary to interpret those signals. This is crucial in developing novel and emerging applications in robotics, self-driving cars, and many more.

© Springer Nature Switzerland AG 2019
B. Moons et al., *Embedded Deep Learning*,
https://doi.org/10.1007/978-3-319-99223-5_1

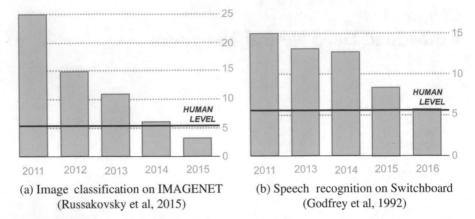

(a) Image classification on IMAGENET
(Russakovsky et al, 2015)

(b) Speech recognition on Switchboard
(Godfrey et al, 1992)

Fig. 1.1 Deep neural networks outperform trained humans on (**a**) 1000-class image recognition and (**b**) real-time speech captioning

Although these networks are extremely powerful, they are also very expensive from a computational and hardware perspective. Most deep learning algorithms are extremely computationally and memory intensive, requiring tens of megabytes for filter coefficient storage and hundreds of millions of operations per input. This high cost makes them difficult to employ on always-on embedded or battery-constrained systems, such as smartphones, smart-glasses, and even autonomous vehicles such as robots, cars, and drones. This book investigates novel applications, adaptations to neural networks, and hardware architectures and circuits that could make this vision of always-on state-of-the-art wearable perceptive capabilities a reality.

This chapter consists of four main parts. Section 1.2 is a quick introduction to some general concepts in machine learning. Section 1.3 is an introduction to deep learning techniques. It discusses artificial neural networks (ANN), convolutional neural networks (CNN), recurrent neural networks (RNN), and how to efficiently train them. Throughout the rest of this text, knowledge of these algorithms is considered a prerequisite. Section 1.4 is a literature overview of the challenges of embedded deep learning on battery-constrained devices and of the currently existing methods and research directions that could make this a reality. Finally, Sect. 1.5 is an overview of my research contributions towards this end-goal of embedded and always-on neural network-based inference.

1.2   Machine Learning

A machine learning algorithm is an algorithm that is able to learn from data (Goodfellow et al. 2016). Here, learning is understood as in Mitchell (1997): "a computer program is said to learn from experience E with respect to some class of tasks T and performance measure P, if its performance at tasks in T, as measured by P, improves with experience E." The following overview is based on Goodfellow et al. (2016), which is an excellent introduction to the field.

1.2.1 Tasks, T

Several tasks T are targeted in machine learning applications. Machine learning is used for regression analyses (Huang et al. 2012), transcription (Shipp et al. 2002), machine translation (Cho et al. 2014; Bahdanau et al. 2014), anomaly detection (Chandola et al. 2009; Erfani et al. 2016), denoising (Vincent et al. 2010), and synthesis techniques (Ze et al. 2013; Rokach et al. 2012). Most important in the context of this dissertation is the **classification** (Glorot et al. 2011; LeCun et al. 2015) task.

In this task, the machine is asked to specify which of k categories some input belongs to. An example of a classification task is object recognition, where the input is an input image and the output is a numeric code identifying the object in the image. Another example is speech recognition, where an input audio waveform is translated into a sequence of classified phonemes or words. Most modern object classification is best accomplished with deep learning algorithms. Deep learning-based classification applications are hence the main scope of this work.

1.2.2 Performance Measures, P

The performance measure P is a quantitative measure of a machine learning algorithm's performance.

P differs depending on the task at hand. For detection, a two-class classification problem, measures for P are **precision** (true returned positives divided by all returned positives) and **recall** (true returned positives divided by all positives). More generally, in classification and transcription tasks, P is the **accuracy** or error rate. Here, accuracy is the proportion of correctly classified examples.

In order to estimate the performance of the machine learning algorithm on data that it has not seen before, as would happen in the real world, it is crucial to verify P on a **test set** of data that is separate from the data used for training the system.

1.2.3 Experience, E

A machine learning algorithm can increase its performance P on a task T by gaining more experience E. Algorithms can be categorized as **supervised** and **unsupervised** depending on the type of E they have access to during their learning process.

1.2.3.1 Supervised Learning

Supervised learning algorithms experience a data set, where each example is associated with a label or target. They learn to **predict** a target **y** from an input **x**. This process is analogous to the concept of a teacher telling a student what to do, hence supervising him to predict the correct result.

1.2.3.2 Unsupervised Learning

Supervised learning algorithms experience a data set and learn useful properties of the structure of this data. They basically model the probability distribution of a property of a data set. In unsupervised learning, no labels or targets are provided, making it a significantly harder task than supervised learning. The algorithm must make sense of the data without any supervision. Examples of algorithms in unsupervised learning are principal component analysis and k-means clustering.

The rest of this text will only focus on algorithms and applications that make use of supervised learning.

1.3 Deep Learning

The rapidly developing field of deep learning is introduced in several excellent works (LeCun et al. 2015; Goodfellow et al. 2016; Li et al. 2016). For decades, conventional machine learning techniques were SotA in object recognition, although they were limited in their ability to process natural data in their raw form. Figure 1.2a illustrates how in conventional machine learning, constructing

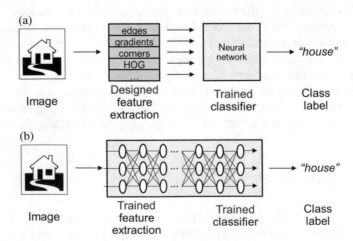

Fig. 1.2 Comparing the (**a**) classical approach to machine learning using hand-designed features to the (**b**) multi-layered representational approach of deep learning operating on raw inputs

a pattern recognition or machine learning system required careful engineering and considerable domain expertise to design a feature extractor that could transform raw data (such as pixel values of an image) into a suitable internal representation or feature vector from which a learning subsystem, typically a classifier, could detect or classify patterns in the input. Examples of such features in the computer vision context can be edges, gradients, specific colors, or more automated forms such as SURF (Bay et al. 2006) or HOG (Dalal and Triggs 2005).

The concept of deep learning breaks with this tradition of using hand-designed features, as it is a type of representation learning (LeCun et al. 2015). Representation learning is a set of methods that allow a machine to be fed with raw data and to automatically discover the representations needed for detection or classification. This is illustrated in Fig. 1.2b, where a raw input image is fed to a deep learning system. In deep learning algorithms, multiple levels of representation are used. They are a multi-layered system composed of nonlinear modules each transforming an input representation (starting with the input images, for example) into an output representation at a higher, more abstract level. If many such layers, or many such transformations, are combined, very complex functions can be learned.

The difference between classical approaches to machine learning and representational deep learning is illustrated in Fig. 1.2. The key aspect of deep learning is that these layers of features are not designed by hand by human engineers: they are learned from data using a general purpose learning procedure, typically a form of stochastic gradient descent (see Sect. 1.3.4). This strategy has been proven to be very powerful and crucial to many of the recent breakthroughs in Machine Learning and pattern recognition, as shown in Fig. 1.1.

Instead of hand-designed by a human expert, the intermediate features generated in representational deep neural network are a mathematical optimum, within the constraints of the optimization method (see Sect. 1.3.4). These features can be visualized through the powerful tools from Olah et al. (2017), which were used to generate Fig. 1.3. Figure 1.3 shows how GoogleNet (Szegedy et al. 2015) builds up its understanding of images over many layers. The first layers extract low-level

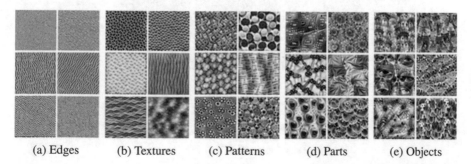

(a) Edges   (b) Textures   (c) Patterns   (d) Parts   (e) Objects

Fig. 1.3 Visualizations of features learned by GoogleNet (Szegedy et al. 2015) on IMA-GENET (Russakovsky et al. 2015), taken from Olah et al. (2017). Learned features vary from simple (**a**) in the first layers to more complex and abstract (**e**) in the final layers

features such as edges and contrast changes, as shown in Fig. 1.3a. Deeper layers use that information to extract more complex features, such as textures (Fig. 1.3b), patterns (Fig. 1.3c), parts (Fig. 1.3d), and ultimately objects (Fig. 1.3e).

The rest of this section introduces and discusses three types of typical neural network architectures that are used in deep learning to perform this automated feature extraction. Basic deep feed-forward neural networks are discussed in Sect. 1.3.1. Convolutional neural networks, their optimized form useful for modeling spatial correlations and leveraging sparse interactions, parameter sharing, and equivariant representations, are discussed in Sect. 1.3.2. Recurrent neural networks, used to model sequential correlations, are discussed in Sect. 1.3.3, although they are somewhat out of scope of this text. Finally, Sect. 1.3.4 dives into some details on how to efficiently train and regularize large-scale neural networks.

1.3.1 Deep Feed-Forward Neural Networks

Deep feed-forward networks are also referred to as multi-layer perceptrons, dense or fully connected networks (FCN), or artificial neural networks (ANN). They are the quintessential deep learning models, which have been around since the 1940s and 1960s. A deep feed-forward neural network has no internal feedback connections, hence its name. A network with a depth of n layers would compute the following function: $f(x) = f^{(n)}(\dots f^{(2)}(f^{(1)}(x)))$, where $f^{(1)}$ is the networks **input** layer, $f^{2\dots n-1}$ are called **hidden** layers, and $f^{(n)}$ is the networks **output** layer. During training, $f(x)$ is trained to come as close as possible to $f^*(x)$, which is the evaluated ground truth for a given problem. These networks are called *neural*, as they are loosely inspired by neuroscience.

Figure 1.4 is an illustration of a three layer **deep** feed-forward neural network. It has one input and output layer and a single hidden layer. The dimension of each of

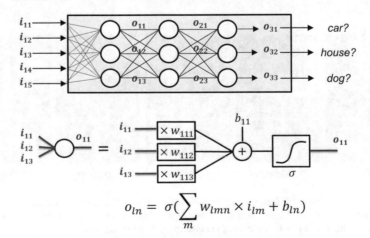

$$o_{ln} = \sigma(\sum_m w_{lmn} \times i_{lm} + b_{ln})$$

Fig. 1.4 Graphical representation of a simple deep feed-forward neural network

the layers is referred to as the **width** of the layer. Here each layer has a width of 3. Each of the units, or neurons, in such a layer performs a dot product or vector-to-scalar function on an input feature vector, generating a single output feature. A full layer contains a vector of neurons and hence performs a vector-to-vector function on an input feature vector x. In order for such networks to model nonlinear functions f, a nonlinear activation function has to be added to the neuron functionality. Typically used options are the sigmoid function $a(z) = \sigma(z)$, the hyperbolic tangent function $a(z) = tanh(z)$, or various forms of rectified linear units (ReLU) $a(z) = max(0, z)$. The functionality of a single neuron can hence be described as in the following equation:

$$O = a \left(\sum_{m=0}^{M} W[m] \times x[m] + B \right) \qquad (1.1)$$

where a can be any of the activation functions listed above. From an algebraic point of view, a single fully connected layer (FCL) can hence be seen as a vector–matrix (input vector x, weight matrix W) product with a per-neuron additive scalar bias B after which an element-wise activation function a is applied. This means FCNs are already supported through heavily optimized BLAS and GEMM libraries for x86 and RISC CPUs.

According to the **universal approximation theorem** (Hornik et al. 1989) the feedforward network above can be proven to be able to approximate any continuous function on a closed and bounded subset of $\mathbb{R}^n$. This is the case only if the network contains at least one hidden layer with *enough* hidden units and if it uses a "squashing" activation function, such as the σ and ReLU functions. Hence, any function we are trying to learn can be represented by a deep feed-forward neural network of sufficient size. However, this does not mean we'll also be able to train this network to map this function. The training algorithm might get stuck in a local minimum or it could converge to the wrong function when it is overfitting. On top of that, deeper networks suffer from *vanishing gradients* (Hochreiter et al. 2001) when plain topologies and optimization techniques are used. This means the gradients for the weights in the first layers become too small for them to be significantly updated. In general, the giant networks that might be necessary to learn complex nonlinear functions are infeasible to train with current technologies. Details on regularization to prevent overfitting and on optimization algorithms are given in Sect. 1.3.4.

1.3.2 Convolutional Neural Networks

Albeit a universal function approximator, deep feed-forward neural networks do suffer from several flaws, mainly linked to the difficulty to train them. Convolutional neural networks (CNN) are a specialized form of deep feed-forward neural networks that try solving these flaws. First, on high-dimensional input data such as images,

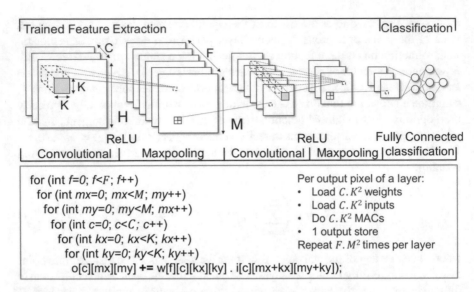

Fig. 1.5 A typical example of a multi-layer convolutional neural network

deep feed-forward input layers will become huge as they are fully connected. For example, a single neuron operating on a VGA image requires almost a million weights. This large amount of parameters leads to overfitting, even when using strong regularizers (Sect. 1.3.4). For many, especially visual applications, this full connectivity is overkill. In visual classification, initially only local connections of $3-10 \times 3-10$ modeling relevant edges are important. In such applications, exploiting **sparse connectivity** is a way to reduce the number of weights in a neural model. On top of that, the same pattern can appear anywhere in an image, therefore it makes sense to use the same filter on multiple position of an input image. This **parameter sharing** is another way to reduce the memory requirements for neural models, and to improve their statistical efficiency, which indicates the number of required operations to achieve a given accuracy.

CNNs, visualized in Fig. 1.5, do exploit the previous characteristics—**sparse connectivity** and **parameter sharing**—to improve the statistical efficiency of neural network models and increase their trainability. These networks are a type of artificial neural networks inspired by visual neuroscience. They are a cascade of multiple stacked convolutional, nonlinearity, and pooling-layers used for feature extraction, followed by a smaller number of fully connected neural network layers used for classification. The number of cascaded stages in recent CNN models varies anywhere from 2 (LeCun et al. 1998), typically 10–20 (Simonyan and Zisserman 2014a) to more than one hundred (He et al. 2016a), ending with typically 1–3 fully connected layers (Krizhevsky et al. 2012a) for classification.

A **convolutional layer** (CONVL), with topology parameters listed in Fig. 1.5 and Table 1.1, transforms input feature maps (I) into output feature maps (O), each containing multiple units. Each unit in an output feature map ($M \times M \times F$) is

Table 1.1 Parameters of a CONV layer

Parameter	Description	Range
F	Number of filters per layer	16–512
H	Width and height of input feature map	16–227
C	Number of channels in input feature map	3–512
K	Width and height of filter plane	1–11
M	Width and height of output feature map	16–227

connected to local patches of units ($K \times K \times C$) in the input feature maps through a filter $W[F]$ ($K \times K \times C \times 1$) in a filter bank W ($K \times K \times C \times F$) existing out of a set of machine-learned weights and a bias (B) per output feature map. A formal mathematical description is given in the following equation:

$$O[f][x][y] = a\left(\sum_{c=0}^{C} \sum_{i=0}^{K} \sum_{j=0}^{K} I[c][Sx + i][Sy + j] \times W[f][c][i][j] \right.$$

$$\left. + B[f] \right)$$

(1.2)

where a is a typical activation function such as ReLU, S is a stride, and x, y, f are bounded by: $x, y \in [0, \ldots, M[$ and $f \in [0, \ldots, F[$. Figure 1.5 shows Eq. (1.2) can be naively implemented as a deep nested loop.

The result of the local sum computed in this filter bank is then passed through a **nonlinearity layer**, typically a ReLU, using the nonlinear activation function $a(z) = max(0, z)$, where u is a feature map unit. This activation function reduces the vanishing gradient problem (Hochreiter et al. 2001) in the backpropagation-based training phase of the network and leads to high degrees of sparsity due to non-activated outputs.

Max-pooling layers compute and output only the maximum of a local patch (typically 2×2 or 3×3) of output units in a feature map. They thereby reduce the dimension of the feature representation and create an invariance to small shifts and distortions in the inputs.

Finally, **fully connected layers** (FC) are used as classifiers in the CNN algorithm. An FC layer is also here described as the matrix–vector product $O[z] = \sum_{m=0}^{M} W[z, m] \times I[m] + B[z]$, equivalent to Eq. (1.1), where M is the size of vectorized input feature map and $z \in [0, \ldots, Z[$ is the number of neurons in the fully connected layer. As all the used weights are only used once in a forward pass, there is no weight reuse in these layers. Due to this observation, architectures proposed for FC layers, as Han et al. (2016) and Reagen et al. (2016), are different from architectures for CONV layers. In general, the optimal network architecture, characterized by the number of cascading stages and the values of model parameters $F, H, C, K,$ and M, varies for each specific application.

Table 1.2 Model size and computational complexity comparison between FC and CONV layers

Network	CONV size [#w]	FC size [#w]	CONV ops [#MAC]	FC ops [#MAC]
LeNet-5 (LeCun et al. 1998)	25.5k	405k	1888k	405k
AlexNet (Krizhevsky et al. 2012a)	2.3M	58.6M	666M	58.6M
VGG-16 (Simonyan and Zisserman 2014a)	14.7M	124M	15.4G	124M
SqueezeNet (Iandola et al. 2016)	733k	0	746M	0

Early top-performing algorithms were AlexNet (Krizhevsky et al. 2012a), a breakthrough implementation with best results on IMAGENET that sparked the current peak in neural network research, and VGG-16 (Simonyan and Zisserman 2014b), which is a deeper version of the same "plain" CNN. Both implementations stick to the same basic network topology discussed here. Both these networks use vast FCNs after their convolutional network, which makes the networks huge and statistically inefficient.

As illustrated in Table 1.2, the total network weight size is dominated by the FC layer weights (>90% of the total network size), while the amount of computations is heavily dominated by the CONV layers (>90% of the total amount of MAC-operations).

Several more advanced forms of convolutional neural networks exist. Most efforts in this research area go to increasing the statistical efficiency of the model topology: hence, minimizing the necessary amount of operations and the memory requirements to achieve a certain nonlinear function. This is discussed in more detail in Sect. 1.4.

Most notable examples are various forms of **residual neural networks** in He et al. (2016b), Zagoruyko and Komodakis (2016), and Xie et al. (2017). Residual neural networks have architectural changes that allow them to be substantially deeper than "plain" networks. In essence, the problem of vanishing gradients (Hochreiter et al. 2001) in "plain" neural networks is minimized by adding various forms of residual connections. The function $f(x)$ is often changed by $f'(x) = f(x) + x$. As these networks can be very deep, they require less weights to achieve the same accuracy compared to their "plain" counterparts. Figure 1.6 shows several examples of more advanced CNN building blocks, among which residual blocks in Fig. 1.6a.

Other notable examples are **inception** networks, first proposed as GoogleNet (Szegedy et al. 2015) and shown in Fig. 1.6b, which were further advanced and embedded in a residual architecture in Szegedy et al. (2017) (Fig. 1.6c). In inception networks, a single layer is split into several parallel layers with different kernel sizes operating on the same input data, as is shown in Fig. 1.6b. This allows modeling spatial and channel-wise correlations as partly decoupled, which makes them easier to train. An extreme form of inception models is **Xception**: where spatial and cross-channel correlations are completely separated (Chollet 2016) by using depthwise separable filters, Fig. 1.7. The resulting architecture slightly outperforms

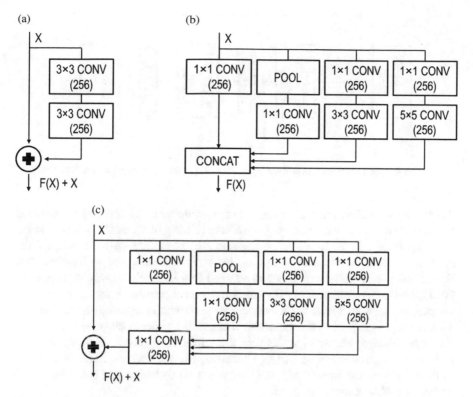

Fig. 1.6 Examples of residual and inception advanced building blocks in modern CNNs. They either improve trainability, allowing for deeper networks, or increase the statistical efficiency or modeling capacity of the network. (**a**) Residual block. (**b**) Inception block. (**c**) Residual inception block

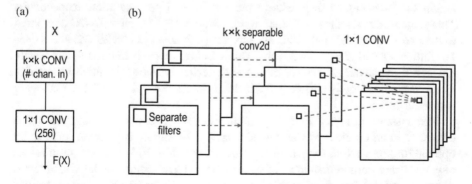

Fig. 1.7 The SotA Xception architecture, used in MobileNets (Howard et al. 2017). (**a**) Xception block. (**b**) Depthwise separable layers

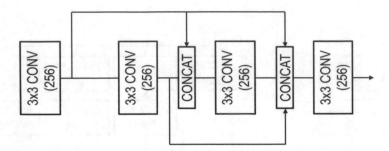

Fig. 1.8 A 4-layer DenseNet block. Each layer takes all preceding feature maps as inputs

the previous inception models using the same number of parameters as shown in Fig. 1.9. Another modern type are **DenseNets** (Huang et al. 2016), which connect the outputs of a layer to all subsequent layers. These are only about 2× more efficient in terms of accuracy/weight on IMAGENET compared to ResNets. Yet, DenseNets are interesting as they can be used in a hierarchical way (Huang et al. 2017), where easy samples are classified early in the network and more difficult samples are classified later in the network as they require more advanced features. This concept is related to the hierarchical cascading concept discussed in Chap. 2. The most efficient networks at this time are **Mobilenets**, based on Xception, also from Google (Howard et al. 2017). Currently, MobileNets are SotA in terms of statistical efficiency: they require the least amount of weights and computations to achieve a usable accuracy (Fig. 1.8).

An interesting analysis of different network architectures is compiled in Figs. 1.9 and 1.10. These are expanded from Canziani et al. (2016) and expanded on Moons et al. (2017d). For a number of recent networks this figure shows the number of computations and weights required as a function of the achieved top-1 accuracy on IMAGENET. Figures 1.11 and 1.12 show the "efficiency" of each of these networks: the accuracy they achieve per weight and per operation, respectively. These figures clearly show the massive improvements of the latest CNNs compared to the first deep models published as AlexNet (Krizhevsky et al. 2012a) and VGG-16 (Simonyan and Zisserman 2014b). This whole analysis is important, as the data flows in these advanced state-of-the-art MobileNets based on depthwise separable filters are considerably different from the flow in the "plain" CNNs that were discussed earlier.

The systems designed in this book in Chaps. 5 and 6 cannot support these dataflows in an efficient way and would have to be significantly redesigned to be optimal for MobileNets, for example. Also note that the CNN-community optimizes only towards a minimum number of operations and a minimum model size in terms of number of weights. A network with a limited number of weights and operations does however not necessarily lead to the most energy-efficient solution. Major parts of the energy consumption on an embedded platform are in on- and off-loading

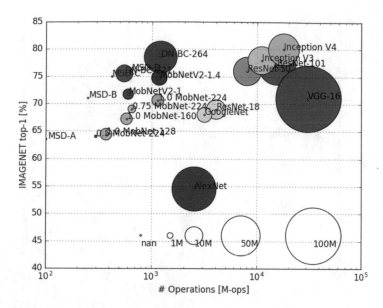

Fig. 1.9 Performance of different network architectures in the accuracy-operations-weights space. MSD are multi-scale DenseNets (Huang et al. 2017), DN are DenseNets (Huang et al. 2016), MobNets are mobilenets (Howard et al. 2017). The size of the blobs indicates the model size in MegaBytes

---

The Bigger Picture

The current performance of neural networks is compared in the three-dimensional ops-weights-accuracy space. A network is deemed superior if it achieves a higher accuracy with the least amount of necessary operations and the least amount of weights. However, this does not necessarily mean the network is faster or consumes less energy than an alternative at the same accuracy. Different network topologies, such as residual networks or depthwise separable filters, might be harder to map on existing hardware than plain networks. On top of that, none of the metrics above take feature map sizes into account, the energy consumption of which might dominate in very wide or deep neural networks.

---

feature maps. If those are large, or if the network is very deep which means it has a lot of intermediate feature maps, the energy of these transfers can dominate. This is not generally acknowledged in the field, but briefly touched upon in Sze et al. (2017) and Moons et al. (2017a). For example, in Sze et al. (2017), it is shown a network that requires 50× less operations than AlexNet (Krizhevsky et al. 2012b) at iso-accuracy consumes 30% more on Eyeriss (Chen et al. 2016) than AlexNet itself. Hence, the field should move towards direct energy metrics, rather than indirect energy metrics such as # operations and # weights that prove to be insufficient measures for network performance.

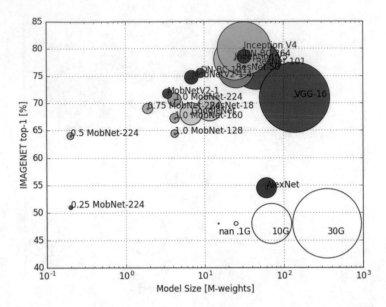

Fig. 1.10 Performance of different network architectures in the accuracy-weights-operations space. MSD are multi-scale DenseNets (Huang et al. 2017), DN are DenseNets (Huang et al. 2016), MobNets are mobilenets (Howard et al. 2017). The size of the blobs indicates the number of necessary operations in Megaflops

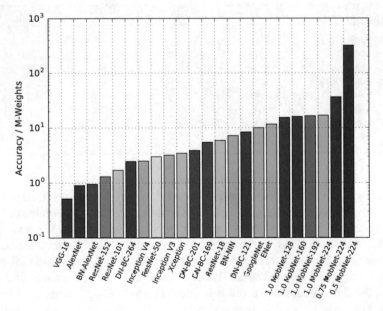

Fig. 1.11 Weight efficiency of different network architectures. MSD are multi-scale DenseNets (Huang et al. 2017), DN are DenseNets (Huang et al. 2016), MobNets are mobilenets (Howard et al. 2017)

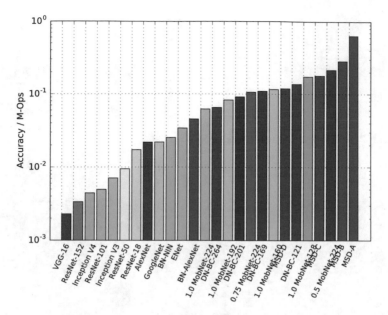

Fig. 1.12 Operations efficiency of different network architectures. MSD are multi-scale DenseNets (Huang et al. 2017), DN are DenseNets (Huang et al. 2016), MobNets are mobilenets (Howard et al. 2017)

1.3.3 Recurrent Neural Networks

Traditional deep feed-forward neural networks have difficulties modeling long-term time dependencies. Recurrent neural networks (RNN) are feed-forward neural networks with internal states and loops, allowing information to persist. This is useful in natural language processing, translations, speech recognition and video processing, for example. Currently, the best performing type of RNN is long short-term memory (LSTM), as it manages to model very long-term time dependencies (Olah 2015).

An excellent introduction to LSTMs can be found online in Olah (2015). The following two figures are taken from this blog. Figure 1.13 shows the basic concept of an RNN. In this figure, the neural network A takes in a feature vector x_t at time t, together with an output c_t which was part of the output of the previous time step. These two inputs are then combined to generate an output vector h_t. Figure 1.13 shows how such networks can be unrolled in time. Figure 1.14 shows how this setup is implemented in an LSTM network. Every LSTM block takes in three vectors: an input x, a state c, and an output h. It then transforms these into a new state and output through applying several LSTM gates, as indicated in Eq. (1.3).

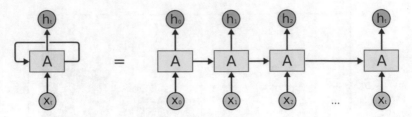

Fig. 1.13 Recurrent neural networks taken from Olah (2015)

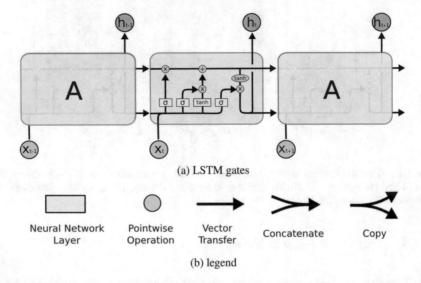

(a) LSTM gates

Neural Network   Pointwise   Vector   Concatenate   Copy
Layer   Operation   Transfer

(b) legend

Fig. 1.14 Operations in an LSTM block, taken from Olah (2015). (**a**) LSTM gates (**b**) legend

$$f_t = \sigma(W_f \cdot [h_{t-1}, x_t] + b_f)$$
$$i_t = \sigma(W_i \cdot [h_{t-1}, x_t] + b_i)$$
$$C_t' = tanh(W_C \cdot [h_{t-1}, x_t] + b_C)$$
$$C_t = f_t \odot C_{t-1} + i_t \odot C_t'$$
$$o_t = \sigma(W_o \cdot [h_{t-1}, x_t] + b_o)$$
$$h_t = o_t \odot tanh(C_t)$$

(1.3)

Here, $\cdot$ is a dot product and $\odot$ is an element-wise multiplication. Hence, from these equations, it is clear that 8 (2×4) deep feed-forward neural network layers are at the core of the LSTM block. Hardware optimized for feed-forward neural layers should hence also be optimized for LSTMs.

1.3.4   *Training Deep Neural Networks*

Neural networks are supervised learning algorithms, meaning they learn to predict targets from an input-label combination. In practice they are trained through different variations of gradient descent optimization. Here all partial derivatives of a **loss** function, a measure for the deviation of the predicted output y' from the ground truth y to the weights in the network, are derived through **backpropagation**. Based on these derivatives, the **optimization algorithm** determines how each of the networks parameters are to be updated in every iteration. This process is performed iteratively on vast data sets of training data. If the data set is too small or not sufficiently diverse, the network might not generalize to model a representative test set and will overfit. A number of **regularization techniques** exist to prevent this. A full discussion is out of scope of this text, but the basic concepts are touched upon in this section. The interested reader is pointed to Goodfellow et al. (2016) for a more elaborate introduction.

1.3.4.1   Loss Functions

In order to optimize a function through gradient-based learning, a differentiable loss function is required. Optimally, the distribution defined by the neural network model should be as close as possible to the distribution of the data that is to be represented. In other words, the model should be optimized for maximum likelihood. Mathematically, this can be enforced through minimizing the cross entropy loss between the training data's labels and the model's predictions. Equivalently (Goodfellow et al. 2016), cross entropy can also be minimized through minimizing the mean squared error (MSE) of the model's predictions.

The categorical cross entropy has proven to be best suited for general, deep neural networks. The loss function and its derivation with respect to the output o_j is given as:

$$L(t,o) = -\sum_{j}^{C} t_j log o_j$$

$$\frac{\partial L}{\partial o_j} = \frac{-t_j}{o_j} \tag{1.4}$$

where C is the number of output neurons, t_j is 1 for the target class and 0 otherwise, and o_j is the system's predicted output value.

The mean squared error is also often used as a loss function. This function and its derivation with respect to the output o_j is given as:

$$L(t, o) = \frac{1}{C} \sum_{j}^{C} \frac{1}{2} (t_j - o_j)^2$$

$$\frac{\partial L}{\partial o_j} = \frac{1}{C} (t_j - o_j) \tag{1.5}$$

Other loss functions exist as well. Hinge-loss, commonly used in quantized neural networks (Chap. 3), is a maximum-margin loss function. Janocha and Czarnecki (2017) is a good overview of the theoretical and practical performance of typical loss functions in deep learning.

1.3.4.2 Backpropagation

Once a good loss function has been chosen, the partial derivatives of this function to any contributing weight or bias in the model can be computed through the chain rule of differentiation. Backpropagation is an efficient algorithm that computes this chain rule with a specific order of operations. This chain rule is given here:

$$\frac{\partial z}{\partial x} = \frac{\partial z}{\partial y} \frac{\partial y}{\partial x} \tag{1.6}$$

The same principle can be used to find the partial derivatives of the loss function to any weight w_{ij} connecting a neuron j to a previous neuron i throughout the model. Let's say o_j is either a softmax or sigmoid activation function transforming the output z_j into output units. The partial derivative of the loss function to that weight is:

$$\frac{\partial L}{\partial w_{ij}} = \frac{\partial L}{\partial o_j} \frac{\partial o_j}{\partial z_j} \frac{\partial z_j}{\partial w_{ij}} \tag{1.7}$$

Or, more generally:

$$\frac{\partial L}{\partial w_{ij}} = \sum_{p} \left[\frac{\partial L}{\partial o_p} \left(\sum_{k} \frac{\partial o_p}{\partial z_k} \frac{\partial z_k}{\partial w_{ij}} \right) \right] \tag{1.8}$$

where $\sum_{p}$ is the summation over all output units and $\sum_{k}$ is the summation over all inputs contributing to o_p. Typically this formula simplifies significantly, as most of the terms in the previous summation will become zero.

1.3.4.3 Optimization

As minimizing the loss function in deep learning is a non convex optimization problem, very robust optimization algorithms are necessary. If not, the optimization might get stuck in one of the many local minima, rather than converge to the global minimum in the loss function. Therefore, neural networks are typically trained through variations on the **stochastic gradient descent** (SGD) algorithm.

Stochastic Gradient Descent

SGD (LeCun et al. 1998) is a stochastic approximation of the baseline gradient descent optimization, in which gradients are estimated through backpropagation based on the loss averaged over m samples of the total data set containing $N >> m$ samples. All parameters θ are then updated via $\theta = \theta - \epsilon g$, where g is the gradient estimate for every parameter and ϵ is a small learning rate. This iterative method has several advantages compared to "plain" gradient descent, where the derivative is computed on the full training set. First, using stochastic gradient descent, the computation time per update does not grow with the number of training examples, but only with the minibatch size m. This is typically small and in the range of 16–256. More importantly, SGD introduces noise, even when close to a minimum. This prevents the algorithm from converging to a local minimum, but might also prevent any convergence at all. In practice, the learning rate is commonly decayed as training progresses to prevent this.

Momentum

is an extension to SGD to accelerate the learning process. Momentum accumulates an exponentially decaying moving average of past gradients and continues to move in their direction. The weight update is now $\theta = \theta + v$, where v is $v = \alpha v - \epsilon g$. Here, $\alpha \in [0, 1]$ is a hyperparameter determining the speed of exponential decay. g is again the gradient estimate computed on a minibatch. If α is larger, the contributions of the previous gradients will be larger. This generally leads to faster convergence times. The momentum vs SGD concept is illustrated in Fig. 1.15.

Fig. 1.15 Performance of momentum (red) versus plain SGD (black). Figure is taken from Goodfellow et al. (2016)

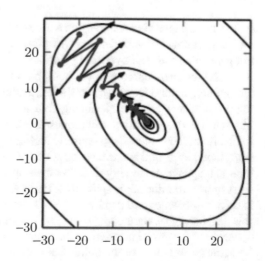

A number of other optimization algorithms exist. Adagrad (Duchi et al. 2011), RMSprop (Tieleman and Hinton 2012), and Adam (Kingma and Ba 2014) are examples with adaptive learning rates. A good overview of the performance of different methods is given in Ruder (2016).

Apart from optimization strategy, also the **initialization** is key. Several initialization strategies exist that are out of the scope of this text.

Another technique in training modern neural networks is batch normalization (Ioffe and Szegedy 2015), which is key to training very deep models. In gradient descent, all parameters are updated under the assumption that all other parameters remain constant. This is of course not the case and might lead to unexpected, unstable results. The effects of an update to the parameters of one layer will strongly depend on all the other layers. In batch normalization, the network is reparametrized to mitigate this problem. In batch normalization a minibatch of feature maps F is reparametrized as F':

$$F' = \frac{F - \mu}{\sigma} \tag{1.9}$$

where μ, σ are the mean and standard deviation of the minibatch feature map. The representational power is then restored through replacing the features by:

$$F'' = \gamma F' + \beta \tag{1.10}$$

where γ and β are trainable scalar parameters. This new parametrization can represent the same functions as before, but has very different, easier learning dynamics.

1.3.4.4   Data Sets

In order to achieve high accuracy on a given task, neural networks have to be trained on a large amount of varied data. If insufficient data is given to the network, it might not generalize and start overfitting. Gathering vast amounts of labeled training data is crucial for deep learning to work. Open-source data sets are abundant in speech recognition, natural language translation, and computer vision. However, in a lot of other fields, they are not. Not being able to gather enough labeled data is a major obstacle for deep learning techniques to be applied outside of its most common fields. As by now, most deep learning technology has been well established, a next step should be to find automated ways of gathering complex data outside of speech, natural language processing (NLP) and computer vision.

A typical training set is split into a **train** and **validation** subset. The train-subset is split into batches and used in the SGD process to minimize the loss function on this data through weight updates. At given intervals, typically an epoch, the loss and accuracy of the model is tested on the validation subset. The validation set is used to track the performance of the training algorithm. It can be used to determine when to stop training or to prevent the network from overfitting. As such, the validation can be seen as a part of the training set, as it is an integral part of the training process.

Ideally, the **test** set is completely independent from the training set. The test set is used to test how the model generalizes to unknown and unseen samples. Hence, the training algorithm should not have seen any samples of the test set during training.

Examples of commonly used data sets throughout this text, and in computer vision in general, are the following:

- **MNIST** (Le Cun et al. 1990) is a small data set of handwritten digits. It contains 70k 28 × 28 grayscale images. This data set is not considered to be relevant to judge the performance of new network architectures, but is rather used as a quick test case in code development. SotA accuracy on this benchmark exceeds 99%
- **CIFAR-10** (Krizhevsky and Hinton 2009) is a small data set containing 10 classes, examples of which are *horse* and *car*, among others. The set contains 60k 32 × 32 RGB color images. Although significantly more complex than MNIST, this data set is mainly used to quickly evaluate new training frameworks or network architectures. A good performance of an algorithm on CIFAR-10 does not necessarily correspond to a good performing system in the wild. The SotA accuracy on this benchmark exceeds 95%.
- **SVHN** (Netzer et al. 2011) is a real-world image data set containing digits and numbers ranging from 0 to 9 in natural scene images. The SVHN is obtained from real house numbers in Google Street View images. The data set consists of a training set of 73k, a test set of 26k, and an extra training set of 531k 32 × 32 color images. In this work, only 30% of the extra training data is used.
- **IMAGENET** (Deng et al. 2009) is a large-scale visual recognition data set containing 1000 classes, among which images of several dog species, goldfish, mountains, cars, and so on. The full data set consists of 150 GB of 256 × 256 RGB images. Because of this, training networks on IMAGENET is a long and tedious task. However, good performance on this data set is generally accepted to be a good measure of a networks performance in the wild.

Other commonly used data sets are CIFAR-100 (Krizhevsky and Hinton 2009), Pascal VOC (Everingham et al. 2010), and COCO (Lin et al. 2014).

1.3.4.5   Regularization

A network is **overfitting**, when it performs well on its training data, but not on new unseen inputs. The concept of under- and overfitting is illustrated in Fig. 1.16. A network is overfitting when a model's **training error** (computed on a training set) is much lower than its **generalization error** (computed on a test or validation set). This is opposite to **underfitting**, when a model is not able to obtain a sufficiently low error value on its training set. Overfitting typically happens when the number of used parameters (hence the capacity of the network) is much larger than the amount of training samples. Underfitting can be solved by increasing the model capacity (by going to deeper and wider networks), which in turn might lead to overfitting.

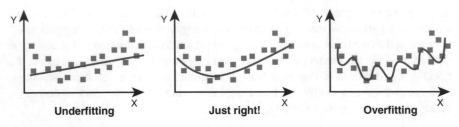

Fig. 1.16 Illustration of under- and overfitting, taken from Gondaliya (2014)

There are a number of effective **regularization** techniques, preventing a model from overfitting:

- **Data augmentation** is an effective means to make a model generalize better. In data augmentation, fake data is generated and added to the training set. In image recognition this is straightforward: flipped, translated, rotated, colorized images of faces are still a representation of the same class. In the real world, the system will encounter such variations naturally. Hence, an augmented data set will be a better representation of the noise omnipresent in the wild. Data augmentation can be applied to speech as well, but not to many other tasks such as natural language processing. If data augmentation does not work, other techniques will have to be applied.

- **Parameter norm penalties** are generally applicable. Here, a penalty is added to the loss function to optimally limit the capacity of a model. The total loss function then becomes $L = L(\theta, y) + \alpha\rho(\theta)$, where $L(\theta, y)$ is the loss function as before, which optimizes accuracy and $\rho(\theta)$ is a factor that only depends on the value of the set of parameters θ. In L^2 regularization or **weight decay**, the weights are driven close to zero by $\rho(\theta) = \frac{1}{2}|w|_2^2$. This will force parameter updates to update in directions which contribute significantly to reducing the objective functions. Parameters corresponding to unimportant directions will be optimized away. In L^1 regularization, $\rho(\theta) = |w|_1$. This regularization will lead to a solution that is more sparse, as a subset of weights will become zero, which can be seen as a feature selection mechanism (Goodfellow et al. 2016).

- **Dropout** (Srivastava et al. 2014) is a regularization method in which some connections are randomly put to zero every iteration. The reader is referred to Goodfellow et al. (2016) for theoretical insight.

- **Adversarial training** (Goodfellow et al. 2014) is a way to prevent overfitting that uses adversarially perturbed images as training samples to increase network accuracy.

1.3.4.6 Training Frameworks

None of the techniques discussed above have to be implemented from scratch. A number of open-source deep learning frameworks exist. These facilitate training different neural network topologies with backpropagation on high-performance computing systems such as CPUs and GPUs. Important factors in deciding which framework to use are (a) the size of its user base, (b) the amount of available code examples and pretrained models, (c) the offered support, and (d) its debugging capabilities. A non-exhaustive list is given below:

- **Tensorflow** (Abadi et al. 2015) is developed and supported by Google. It is used by an enormous community of experts, which provide open-sourced implementations and pretrained models of SotA networks. Keras (Chollet et al. 2015) is a tensorflow wrapper that provides support for fast data loading, data augmentation, and so on. Tensorflow's (a) user base is huge, (b) a lot of code is readily available, (c) the framework is well documented and supported through Google, but is (d) difficult to debug as all network graphs are static and precompiled, not on the fly (as in default python). This means calculations will be fast, but debugging might be harsh, although there is some support.
- **Pytorch** (Paszke et al. 2017) and Torch are open-source frameworks based on Lua, with widespread user bases. As Pytorch is pushed by Facebook, it has a growing user base and supports (a,b,c). In Pytorch, graphs are dynamic and compiled on the fly, which makes them slower, but easier to debug (d).
- **Theano** (Theano Development Team 2016) is an academic project similar in nature to Tensorflow, also using static graphs (d). Lasagna (Dieleman ct al. 2015) is a wrapper similar to Kcras, specifically for Theano. As Theano has been discontinued in 2017, it scores poorly on (a,b,c). It is still listed here, as it was used for several projects throughout this book.
- **Caffe** (Jia et al. 2014) is an academic project with static graphs (d) that has evolved into Caffe2. As Caffe was one of the earliest frameworks out there, it built a large user base quickly (a,b,c). Caffe is also listed here as it was used for experiments throughout this text.

In general, the used deep learning frameworks vary quickly, although Tensorflow at least seems here to stay. Throughout the course of this work, I've moved from Caffe to Theano and Tensorflow in 3 years. This fluidity holds back some industries from adopting any of the frameworks, as they cannot be sure their investments in developing will pay off. Today, some industry efforts are in standardizing deep learning frameworks, which should alleviate these problems.

1.4 Challenges for Embedded Deep Neural Networks

As discussed above, deep learning techniques have become immensely popular for image recognition, as well as for other recognition and pattern matching tasks. Both the training of a deep network and its inference to perform new classifications are currently typically executed on power-hungry CPU servers or GPUs (Fig. 1.17a) in the *cloud*. This setup poses a number of problems for wearable devices. First, it leads to latency and connectivity issues, as a wearable application should always be connected to the cloud. Second, such a setup leads to privacy concerns, as users have to share raw data such as images and speech with a provider. Finally, the necessary

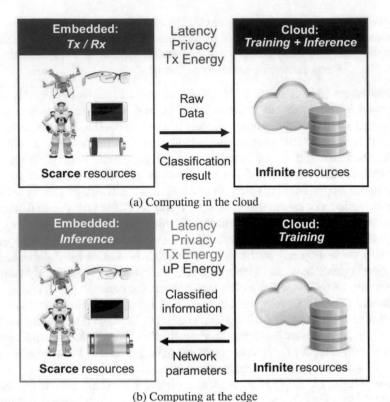

(a) Computing in the cloud

(b) Computing at the edge

Fig. 1.17 (**a**) Current cloud-computing-based embedding of deep learning applications on wearable platforms. (**b**) Future edge-computing-based embedding of deep learning applications on wearable platforms

wireless connections to the cloud are very inefficient and will drain a battery quickly. Because of these reasons, there is a strong demand to move especially the inference step out of the cloud into the *edge*: mobiles, wearables, and IoT sensing nodes. This setup would alleviate latency and privacy issues (Fig. 1.17) and would improve battery life if sufficiently energy-efficient edge processing platforms can be developed.

Current commercial devices lack the capabilities to enable deep inference for real life applications, as illustrated in Fig. 1.18. Recent real-time neural networks for image or speech processing easily require more than 100GOPS to 1TOPS, while requiring the fetching of millions of network parameters (kernel weights and biases) per network evaluation. The energy consumed in these numerous operations and data fetches is the main bottleneck for embedded inference in energy scarce mWatt or uWatt devices. Currently microcontrollers and embedded GPUs are limited to efficiencies of a few 10s to 100s GOPS/W, while embedded always-on inference will only truly be enabled with system-level efficiencies well beyond 10 16b-TOPS/W. To this end, several improved GPUs and neural processing units (NPU)

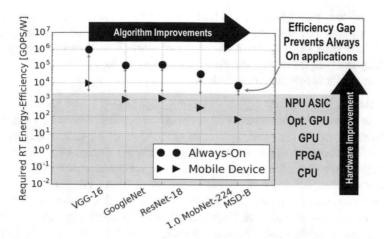

Fig. 1.18 Required full system equivalent energy-efficiency when running several state-of-the-art neural networks with similar IMAGENET top-1 accuracy (69–71%) in real-time (30 fps). Even though new networks have a higher statistical efficiency, hardware innovations, in the form of optimized GPU's and NPU ASICs, are required to close the energy-efficiency gap, especially for always-on applications with a 1 mW power budget. Most recent innovations do bring hardware platforms within the 100 mW mobile device power budget. Finding ways to increase the energy-efficiency has been a major part of this book

ASICs have recently been introduced. However, even though these systems are now theoretically in range of mobile smartphone applications, they are still not efficient enough to bring real-time neural network processing within the always-on power envelope of about 1 mW. Figure 1.18 shows this more clearly. The figure shows several generations of iso-accuracy network architectures and their required energy-efficiency for different power budgets for real-time operation at 30 fps. Although there have been both algorithmic improvements and hardware improvements over the last couple of years, there does remain a large efficiency gap between application requirements and what SotA platforms can offer.

However, overcoming the processing and memory bottleneck in neural networks for always-on applications is possible. This requires a tight interplay between applications, algorithmic optimizations (modifying the network topology), and hardware optimization (modifying the processing architectures and circuits). Finding ways to do this is the main topic of this dissertation.

1.5   Book Contributions

In order to make embedded deep neural network inference a reality, significant innovations will have to be made on all levels of an embedded system: on the application- and algorithmic level, as well as on the processor architecture- and on the circuit-level. Throughout my PhD, I have contributed to the SotA on these three levels—applications, algorithms, and circuits—through various research projects. I hope these contributions can help make the promise of always-on deep neural network processing on battery-constrained devices a reality.

The structure of this book is as follows:

- Chapter 2 discusses how typical detection applications based on deep learning can be optimized to consume as little energy as possible without sacrificing any performance P on their task T. To this end, the general concept of hierarchical cascaded processing is introduced and applied in the context of a 100-class face recognition system, showing up to $100\times$ improvement over a wake-up-based scenario. Goetschalckx et al. (2018) is my contribution on this subject.
- Chapter 3 discusses techniques for several hardware-algorithm co-optimizations that reduce the energy consumption of deep learning algorithms, both by compressing models and by reducing their arithmetic complexity. It is shown that the energy consumption in a neural network accelerator can be greatly reduced by using quantized neural networks—networks with all weights and activations constrained to specific values at train time. The chapter further discusses how hardware and algorithm can be co-designed towards minimal energy consumption. Moons et al. (2016, 2017a) are my contributions on this subject.
- Chapter 4 discusses how embedded deep learning can be made more energy-efficient through various means of approximate computing. This is a set of primarily circuit techniques that allow trading-off energy for computational accuracy whenever the application allows this. This section discusses several state-of-the-art techniques that can achieve this: from simplified error-prone digital arithmetic blocks to voltage over-scaling techniques and analog computational blocks. As most of these techniques are static—their trade-off is fixed at design time—they cannot be used in the context of QNNs and hierarchical cascaded processing, as discussed in Chaps. 2 and 3. Therefore, dynamic-voltage-accuracy-frequency-scaling (DVAFS) is introduced, a dynamic means of approximate computing, with the largest energy-accuracy trade-off shown in the related literature. Moons and Verhelst (2015, 2016, 2017), and Moons et al. (2017b,c) discuss my contribution in terms of the block- and system-level performance of DVAFS, as well as examples of silicon implementations.

The Bigger Picture

In order to bring neural network-based applications to embedded mobile and always-on devices, co-optimizations are necessary across three levels in their design hierarchy. Designers should focus on hardware aware innovations on the application level (Chap. 2), the algorithmic level (Chap. 3), and on the hardware- and circuit-level itself (Chaps. 4, 5, and 6).

- The techniques from Chaps. 2, 3, and 4 are brought together in a physical implementation in Chap. 5. It discusses the architectural design and IC implementation of Envision: a family of dynamic-voltage-accuracy-frequency-scalable (DVAFS) convolutional neural network processors with varying precision, optimized for real-time, large-scale hierarchical processing. These chips were first presented in Moons and Verhelst (2016, 2017), and Moons et al. (2017c).
- The Envision chips are efficient, but not sufficient for always-on wake-up processing. They can hence not be used as an always-on detector as a wake-up stage of a hierarchical application pipeline, discussed in Chap. 2. Chapter 6 discusses BinarEye: a family of always-on binary convolutional neural network processors with all memory on chip that can be used as a visual wake-up sensor. BinarEye leverages two key ideas to minimize its power consumption and energy per classification. First, its computations are in-memory: a massively parallel architectural technique that minimizes data movement and hence minimizes energy consumption. Second, a BinaryNet is an extreme form of a QNN, where arithmetic complexity is minimized and networks are maximally compressed, albeit at a slight accuracy loss. This allows computations in the more efficient analog domain. BinarEye exists in two forms: all-digital BinarEye and mixed-signal MSBNN: mixed-signal binary neural network. Apart from an efficient neural network implementation, this project is hence also a case study in comparing digital and analog compute. MSBNN and BinarEye were presented in Bankman et al. (2018) and Moons et al. (2018), respectively. Follow-up work on advanced voltage scaling (Yang et al. 2018) and CNFET implementations (Hills et al. 2018).
- The final Chap. 7 of this work provides an overview of the realized work, as well as extensive conclusions, trends, and future work required to make the vision of always-on embedded neural networks on battery-constrained devices a reality.

References

Abadi M, Agarwal A, Barham P, Brevdo E, Chen Z, Citro C, Corrado GS, Davis A, Dean J, Devin M, Ghemawat S, Goodfellow I, Harp A, Irving G, Isard M, Jia Y, Jozefowicz R, Kaiser L, Kudlur M, Levenberg J, Mané D, Monga R, Moore S, Murray D, Olah C, Schuster M, Shlens J, Steiner B, Sutskever I, Talwar K, Tucker P, Vanhoucke V, Vasudevan V, Viégas F, Vinyals O, Warden P, Wattenberg M, Wicke M, Yu Y, Zheng X (2015) TensorFlow: large-scale machine learning on heterogeneous systems. https://www.tensorflow.org/, software available from tensorflow.org

Bahdanau D, Cho K, Bengio Y (2014) Neural machine translation by jointly learning to align and translate. Preprint arXiv:14090473

Bankman D, Yang L, Moons B, Verhelst M, Murmann B (2018) An always-on 3.8$u m u$j/classification 86accelerator with all memory on chip in 28nm CMOS. ISSCC technical digest

Bay H, Tuytelaars T, Van Gool L (2006) Surf: speeded up robust features. In: Computer vision–ECCV 2006, pp 404–417

Canziani A, Paszke A, Culurciello E (2016) An analysis of deep neural network models for practical applications. Preprint arXiv:160507678

Chandola V, Banerjee A, Kumar V (2009) Anomaly detection: a survey. ACM Comput Surv 41(3):15

Chen YH, Krishna T, Emer J, Sze V (2016) Eyeriss: an energy-efficient reconfigurable accelerator for deep convolutional neural networks. ISSCC Dig of Technical papers, pp 262–263

Chiu CC, Sainath TN, Wu Y, Prabhavalkar R, Nguyen P, Chen Z, Kannan A, Weiss RJ, Rao K, Gonina K, et al (2017) State-of-the-art speech recognition with sequence-to-sequence models. Preprint arXiv:171201769

Cho K, Van Merriënboer B, Gulcehre C, Bahdanau D, Bougares F, Schwenk H, Bengio Y (2014) Learning phrase representations using RNN encoder-decoder for statistical machine translation. P arXiv:14061078

Chollet F (2016) Xception: deep learning with depthwise separable convolutions. Preprint arXiv:161002357

Chollet F, et al (2015) Keras. https://github.com/keras-team/keras

Dalal N, Triggs B (2005) Histograms of oriented gradients for human detection. In: IEEE computer society conference on computer vision and pattern recognition, 2005. CVPR 2005, vol 1. IEEE, New York, pp 886–893

Deng J, Dong W, Socher R, Li LJ, Li K, Fei-Fei L (2009) Imagenet: a large-scale hierarchical image database. In: Proceedings of the IEEE conference on computer vision and pattern recognition (CVPR), pp 248–255

Dieleman S, Schlüter J, Raffel C, Olson E, Sønderby SK, Nouri D, et al (2015) Lasagne: First release. http://dx.doi.org/10.5281/zenodo.27878

Duchi J, Hazan E, Singer Y (2011) Adaptive subgradient methods for online learning and stochastic optimization. J Mach Learn Res 12:2121–2159

Erfani SM, Rajasegarar S, Karunasekera S, Leckie C (2016) High-dimensional and large-scale anomaly detection using a linear one-class SVM with deep learning. Pattern Recogn 58:121–134

Esteva A, Kuprel B, Novoa RA, Ko J, Swetter SM, Blau HM, Thrun S (2017) Dermatologist-level classification of skin cancer with deep neural networks. Nature 542(7639):115–118

Everingham M, Van Gool L, Williams CK, Winn J, Zisserman A (2010) The pascal visual object classes (VOC) challenge. Int J Comput Vis 88(2):303–338

Glorot X, Bordes A, Bengio Y (2011) Domain adaptation for large-scale sentiment classification: a deep learning approach. In: Proceedings of the 28th international conference on machine learning (ICML-11), pp 513–520

Godfrey JJ, Holliman EC, McDaniel J (1992) Switchboard: telephone speech corpus for research and development. In: IEEE international conference on acoustics, speech, and signal processing, 1992. ICASSP-92, 1992, vol 1. IEEE, New York, pp 517–520

Goetschalckx K, Moons B, Lauwereins S, Andraud M, Verhelst M (2018) Optimized hierarchical cascaded processing. IEEE J Emerging Sel Top Circuits Syst. https://doi.org/10.1109/JETCAS. 2018.2839347

Gondaliya A (2014) Regularization implementation in R: bias and variance diagnosis. http:// pingax.com/regularization-implementation-r/s. Accessed 2018-02-01

Goodfellow I, Pouget-Abadie J, Mirza M, Xu B, Warde-Farley D, Ozair S, Courville A, Bengio Y (2014) Generative adversarial nets. In: Advances in neural information processing systems, pp 2672–2680

Goodfellow I, Bengio Y, Courville A (2016) Deep learning. MIT Press, Cambridge

Han S, Liu X, Mao H, Pu J, Pedram A, Horowitz MA, Dally WJ (2016) EIE: efficient inference engine on compressed deep neural network. In: International symposium on computer architecture (ISCA)

He K, Zhang X, Ren S, Sun J (2016a) Deep residual learning for image recognition. In: Proceedings of the IEEE conference on computer vision and pattern recognition (CVPR)

He K, Zhang X, Ren S, Sun J (2016b) Deep residual learning for image recognition. In: Conference on computer vision and pattern recognition (CVPR)

Hills G, Park R, Shulaker M, Hillard J, Kahng A, Wong S, Bankman D, Moons B, Yang L, Verhelst M, Murmann B, Mitra S (2018) Trig: hardware accelerator for inference-based applications and experimental demonstration using carbon nanotube FETs. In: Design automation conference (DAC)

Hochreiter S, Bengio Y, Frasconi P, Schmidhuber J (2001) Gradient flow in recurrent nets: the difficulty of learning long-term dependencies

Hornik K, Stinchcombe M, White H (1989) Multilayer feedforward networks are universal approximators. Neural Netw 2(5):359–366

Howard AG, Zhu M, Chen B, Kalenichenko D, Wang W, Weyand T, Andreetto M, Adam H (2017) Mobilenets: efficient convolutional neural networks for mobile vision applications. Preprint arXiv:170404861

Huang G, Liu Z, Weinberger KQ, van der Maaten L (2016) Densely connected convolutional networks. Preprint arXiv:160806993

Huang G, Chen D, Li T, Wu F, van der Maaten L, Weinberger KQ (2017) Multi-scale dense convolutional networks for efficient prediction. Preprint arXiv:170309844

Huang GB, Zhou H, Ding X, Zhang R (2012) Extreme learning machine for regression and multiclass classification. IEEE Trans Syst Man Cybern Part B 42(2):513–529

Iandola FN, Moskewicz MW, Ashraf K, Han S, Dally WJ, Keutzer K (2016) Squeezenet: alexnet-level accuracy with 50x fewer parameters and <1mb model size. CoRR abs/1602.07360

Ioffe S, Szegedy C (2015) Batch normalization: accelerating deep network training by reducing internal covariate shift. Preprint:150203167

Janocha K, Czarnecki WM (2017) On loss functions for deep neural networks in classification. Preprint arXiv:170205659

Jia Y, Shelhamer E, Donahue J, Karayev S, Long J, Girshick R, Guadarrama S, Darrell T (2014) Caffe: convolutional architecture for fast feature embedding. Preprint arXiv:14085093

Kingma D, Ba J (2014) Adam: a method for stochastic optimization. ArXiv preprint:14126980

Krizhevsky A, Hinton G (2009) Learning multiple layers of features from tiny images. Technical report

Krizhevsky A, Sutskever I, Hinton GE (2012a) Imagenet classification with deep convolutional neural networks. In: Proceedings of advances in neural information processing systems, pp 1097–1105

Krizhevsky A, Sutskever I, Hinton GE (2012b) ImageNet Classification with Deep Convolutional Neural Networks. In: Pereira F, Burges CJC, Bottou L, Weinberger KQ (eds) Advances in neural information processing systems, vol 25. Curran Associates, Inc., Red Hook, pp 1097–1105. http://papers.nips.cc/paper/4824-imagenet-classification-with-deep-convolutional-neural-networks.pdf

Le Cun BB, Denker JS, Henderson D, Howard RE, Hubbard W, Jackel LD (1990) Handwritten digit recognition with a back-propagation network. In: Advances in neural information processing systems, Citeseer

LeCun Y, Bottou L, Bengio Y, Haffner P (1998) Gradient-based learning applied to document recognition. Proc IEEE 86(11):2278–2234

LeCun Y, Bengio Y, Hinton G (2015) Deep learning. Nature 521(7553):436–444

Li FF, et al (2016) CS231n: convolutional neural networks for visual recognition. http://cs231n.github.io/. Accessed 10 Oct 2017

Lin TY, Maire M, Belongie S, Hays J, Perona P, Ramanan D, Dollár P, Zitnick CL (2014) Microsoft COCO: common objects in context. In: European conference on computer vision. Springer, Berlin, pp 740–755

Mitchell TM (1997) Machine learning, vol 45(37). McGraw Hill, Burr ridge, IL pp 870–877

Moons B, Verhelst M (2015) DVAS: dynamic voltage accuracy scaling for increased energy-efficiency in approximate computing. In: International symposium on low power electronics and design (ISLPED). https://doi.org/10.1109/ISLPED.2015.7273520

Moons B, Verhelst M (2016) A 0.3-2.6 tops/w precision-scalable processor for real-time large-scale convnets. In: Proceedings of the IEEE symposium on VLSI circuits, pp 178–179

Moons B, Verhelst M (2017) An energy-efficient precision-scalable convnet processor in 40-nm cmos. IEEE J Solid State Circuits 52(4):903–914

Moons B, De Brabandere B, Van Gool L, Verhelst M (2016) Energy-efficient convnets through approximate computing. In: Proceedings of the IEEE winter conference on applications of computer vision (WACV), pp 1–8

Moons B, Goetschalckx K, Van Berckelaer N, Verhelst M (2017a) Minimum energy quantized neural networks. In: Asilomar conference on signals, systems and computers

Moons B, Uytterhoeven R, Dehaene W, Verhelst M (2017b) DVAFS: Trading computational accuracy for energy through dynamic-voltage-accuracy-frequency-scaling. In: 2017 design, automation & test in Europe conference & exhibition (DATE). IEEE, New York, pp 488–493

Moons B, Uytterhoeven R, Dehaene W, Verhelst M (2017c) Envision: a 0.26-to-10 tops/w subword-parallel dynamic-voltage-accuracy-frequency-scalable convolutional neural network processor in 28nm FDSOI. In: International solid-state circuits conference (ISSCC)

Moons B, et al (2017d) Bertmoons github page. http://github.com/BertMoons. Accessed 01 Jan 2018

Moons B, Bankman D, Yang L, Murmann B, Verhelst M (2018) Binareye: an always-on energy-accuracy-scalable binary CNN processor with all memory on-chip in 28nm CMOS. In: IEEE custom integrated circuits conference (CICC)

Netzer Y, Wang T, Coates A, Bissacco A, Wu B, Ng AY (2011) Reading digits in natural images with unsupervised feature learning. In: NIPS workshop

Olah C (2015) Understanding LSTM networks. http://colah.github.io/posts/2015-08-Understanding-LSTMs. Accessed 2018-02-01

Olah C, Mordvintsev A, Schubert L (2017) Feature visualization. Distill https://doi.org/10.23915/distill.00007. https://distill.pub/2017/feature-visualization

Paszke A, Gross S, Chintala S, Chanan G, Yang E, DeVito Z, Lin Z, Desmaison A, Antiga L, Lerer A (2017) Automatic differentiation in Pytorch

Reagen B, Whatmough P, Adolf R, Rama S, Lee H, Lee SK, Hernandez-Lobato JM, Wei GY, Brooks D (2016) Minerva: enabling low-power, highly-accurate deep neural network accelerators. In: Proceedings of the ACM/IEEE 43rd annual international symposium on computer architecture (ISCA)

Rokach L, Feldman A, Kalech M, Provan G (2012) Machine-learning-based circuit synthesis. In: IEEE 27th Convention of Electrical & Electronics Engineers in Israel (IEEEI), 2012. IEEE, New York, pp 1–5

Ruder S (2016) An overview of gradient descent optimization algorithms. Preprint arXiv:160904747

Russakovsky O, Deng J, Su H, Krause J, Satheesh S, Ma S, Huang Z, Karpathy A, Khosla A, Bernstein M, et al (2015) Imagenet large scale visual recognition challenge. Int J Comput Vis 115(3):211–252

Shipp MA, Ross KN, Tamayo P, Weng AP, Kutok JL, Aguiar RC, Gaasenbeek M, Angelo M, Reich M, Pinkus GS, et al (2002) Diffuse large b-cell lymphoma outcome prediction by gene-expression profiling and supervised machine learning. Nat Med 8(1):68

Simonyan K, Zisserman A (2014a) Very deep convolutional networks for large-scale image recognition. CoRR abs/1409.1556

Simonyan K, Zisserman A (2014b) Very deep convolutional networks for large-scale image recognition. CoRR abs/1409.1556

Srivastava N, Hinton GE, Krizhevsky A, Sutskever I, Salakhutdinov R (2014) Dropout: a simple way to prevent neural networks from overfitting. J Mach Learn Res 15(1):1929–1958

Sze V, Yang TJ, Chen YH (2017) Designing energy-efficient convolutional neural networks using energy-aware pruning. CVPR

Szegedy C, Liu W, Jia Y, Sermanet P, Reed S, Anguelov D, Erhan D, Vanhoucke V, Rabinovich A (2015) Going deeper with convolutions. In: Proceedings of the IEEE conference on computer vision and pattern recognition, pp 1–9

Szegedy C, Ioffe S, Vanhoucke V, Alemi AA (2017) Inception-v4, inception-resnet and the impact of residual connections on learning. In: AAAI, pp 4278–4284

Theano Development Team (2016) Theano: a Python framework for fast computation of mathematical expressions. arXiv e-prints abs/1605.02688, http://arxiv.org/abs/1605.02688

Tieleman T, Hinton G (2012) Rmsprop: Divide the gradient by a running average of its recent magnitude. coursera: neural networks for machine learning. Technical report

Van Keirsbilck M, Moons B, Verhelst M (2018) Resource aware design of a deep convolutional-recurrent neural network for speech recognition through audio-visual sensor fusion. Arxiv

Vincent P, Larochelle H, Lajoie I, Bengio Y, Manzagol PA (2010) Stacked denoising autoencoders: learning useful representations in a deep network with a local denoising criterion. J Mach Learn Res 11:3371–3408

Xie S, Girshick R, Dollár P, Tu Z, He K (2017) Aggregated residual transformations for deep neural networks. In: 2017 IEEE conference on computer vision and pattern recognition (CVPR). IEEE, New York, pp 5987–5995

Yang L, Bankman D, Moons B, Verhelst M, Murmann B (2018) Bit error tolerance of a CIFAR-10 binarized convolutional neural network processor. In: IEEE international symposium on circuits and systems (ISCAS)

Zagoruyko S, Komodakis N (2016) Wide residual networks. Preprint arXiv:160507146

Ze H, Senior A, Schuster M (2013) Statistical parametric speech synthesis using deep neural networks. In: 2013 IEEE international conference on acoustics, speech and signal processing (ICASSP). IEEE, New York, pp 7962–7966

Chapter 2
Optimized Hierarchical Cascaded Processing

This chapter is largely based on its related publication (Goetschalckx et al. 2018).

2.1 Introduction

The energy bottleneck preventing neural networks from being embedded on mobile and always-on devices, as discussed in Chap. 1, can be solved by developing more energy-efficient algorithms (Han et al. 2015; Liu and Deng 2017; Huang et al. 2017; Moons et al. 2017a) (see Chap. 3), by developing more energy-efficient hardware (see Chaps. 4, 5 and 6) and by developing more intelligent systems that can exploit application-level characteristics to increase energy efficiency. This chapter focuses on the latter approach: increasing system-level energy efficiency through a hierarchical cascaded system that exploits input data statistics.

Some contemporary state-of-the-art (SotA) detection or classification systems reduce power consumption of always-on systems by adding a wake-up stage. This strategy is typically applied in video surveillance systems (Yuan et al. 2009) and key-word recognition (Sun et al. 2017) or voice-activation (Badami et al. 2015; Price et al. 2017). Here, a first stage performs a simple task with a low cost instead of powering on the full, more expensive functionality constantly. This reduces the average on-time of a more expensive classifier considerably and hence the global system's energy consumption. These systems typically do not expand upon this approach outside of a two-stage hierarchy, making them sub-optimal in many cases. Other alternatives try minimizing energy consumption and maximizing performance through building single-function cascades for binary (Viola and Jones 2001; Saberian and Vasconcelos 2014; Panda et al. 2015) or multi-class problems (Venkataramani et al. 2015), or by building tree-based structures that perform multi-class recognition tasks (Venkataramani et al. 2015; Li et al. 2015; Xu et al. 2014; Ghahramani et al. 2010). However, the previous systems generally

© Springer Nature Switzerland AG 2019
B. Moons et al., *Embedded Deep Learning*,
https://doi.org/10.1007/978-3-319-99223-5_2

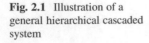

Fig. 2.1 Illustration of a general hierarchical cascaded system

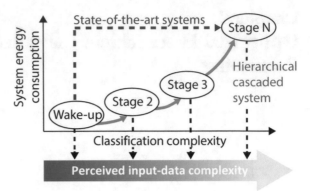

do not explicitly exploit input data characteristics. Furthermore, there exists no framework that can be used to optimize them towards minimum energy consumption at a given accuracy which is useful in the context of embedded processing on battery-constrained devices. In this chapter, cascaded systems are generalized into a full-blown multi-level hierarchical cascade that outperforms the wake-up approach by up to two orders of magnitude at iso-accuracy, as it more adequately exploits data statistics.

An introduction to such a system is given in Fig. 2.1. Here, multiple hierarchical stages are cascaded and can potentially filter out data early. In this setup, both the complexity of the sub-task and the cost of the sub-task increase further down the hierarchical chain. Common classes such as "silence," "Alexa," or "you" in the speech-recognition context are recognized early in the classification hierarchy with inexpensive classifiers, potentially preventing the final expensive stages from being powered on. Hence, even though later stages are more expensive, they are not used as much due to the discriminative functionality of the earlier stages. This proposed architecture combines the benefits of cascaded and tree-based topologies. As in the linear cascade approach, most negative samples are eliminated before the last stage(s), saving power in the overall system. As in a tree-based approach, multi-class problems are supported. Yet, early mis-detections can still be corrected in later stages. This work provides a framework used to minimize overall energy consumption or computational cost in these hierarchical cascades, while simultaneously maximizing or maintaining system-level accuracy.

In building such a framework, this chapter offers the following key contributions:

- It **generalizes wake-up systems to hierarchical cascaded classifiers**: a sequence of increasingly complex classifiers building up to a final multi-class classification problem.
- It **proposes a theoretical roofline model** to gain insight in the performance of a stage in such a hierarchical cascade.
- It **derives general trade-offs** in a generic hierarchical model. Both the impact of input data statistics, the number of stages used, and the individual performance of stages in the hierarchy are discussed.

- It **validates** the hierarchical design approach in a visual recognition case study. This system can dynamically trade energy versus quality by changing the cascade's hyperparameters.

The full chapter is organized as follows. Section 2.2 discusses the novel hierarchical cascaded approach. Both the relevant terminology and the final optimization problem are discussed. Section 2.3 discusses the impact of several data- and system-level parameters on a generic hierarchical system. Section 2.4 applies the proposed theory to a 100-face recognition application. Finally, Sect. 2.5 concludes this chapter.

2.2 Hierarchical Cascaded Systems

This section introduces hierarchical cascaded systems: a multi-stage generalization of wake-up-based systems.

2.2.1 Generalizing Two-Stage Wake-Up Systems

In hierarchical cascaded systems, the overall computational cost of a classification system is minimized without sacrificing performance. This is done by building a functional hierarchy optimized for system-level energy efficiency: separate blocks with increasing functionality and cost are concatenated and jointly optimized. The hierarchy is more efficient than a single stage system, if the early stages are cheap, yet filter data adequately towards the more expensive later stages, without making too many mistakes that cannot be recovered.

The overall system is divided into N stages, as depicted in Fig. 2.2. The end task of the system is assumed to be a complex multi-class classification task. For instance, 100 faces in an image recognition application. The first task typically is a binary wake-up detector.

A typical hierarchy starts with a simple binary classification that removes the most obvious negative samples, such as background images or acoustic noise. If an input is classified as a positive sample, for example, a meaningful image, the next stage is activated to make a more precise classification. Only samples that are detected as positives by the previous stage are fed to the following stage, as is the case in a classical cascade. Throughout this text, the positive class in this framework is called "pass-on-class" (poc) and is shown in Fig. 2.2. However, unlike previous works, every next stage performs a more complex classification task. With this higher complexity and performance, its cost increases in a very nonlinear way.

In order to build a hierarchical cascaded processing system, several alternatives at varying cost-accuracy trade-offs are built and trained for each stage of the hierarchy. Every individual stage is trained with a data set modeling its specific sub-task of the

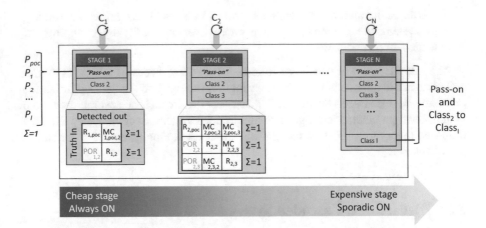

Fig. 2.2 Basic hierarchical classifier system with N stages and I classes $\in [poc, 2, .., I]$. The a-priori chance of each class is P_i. Each stage n has an energy cost of C_i and is described through its pass-on-rate for every class POR_i, its recall R_i, and its misclassification rate $MC_{i,j}$. The final per class recall is R_i as calculated in Eq. (2.5)

full-scale problem. From the system point of view, a stage is then abstracted by its performance (confusion matrix, ROC curve, etc.) and the whole abstracted system is then automatically optimized in order to achieve a given accuracy or recall at minimal complexity or cost. Details on this optimization problem are provided in Sect. 2.2.4.

2.2.2   Hierarchical Cost, Precision, and Recall

In order to automatically optimize hierarchical cascaded systems, descriptions of the system's total cost and recall are derived. The performance of each stage n in Fig. 2.2 can be described using three separate entities. A stage n is defined by specifying its cost C_n and 2 per-class entities: recall $R_{n,i}$, pass-on-rate $POR_{n,i}$. Apart from these factors, the system's total cost $C_{1 \rightarrow N}$ is also heavily influenced by the statistics of the input data, described by the a-priori probabilities of occurrence P_i of each class (Fig. 2.2). Based on all these terms, the overall system recall and cost can be derived.

Here, recall $R_{n,i}$ is the probability that an instance of a given class i is classified correctly in stage n,

$$R_{n,i} = P(\hat{y}_n = i | y = i) \tag{2.1}$$

which is according to its standard multi-class definition. The pass-on-rate $POR_{n,i}$ is the probability that an instance of class i gets passed along to the next classifier by stage n

$$POR_{n,i} = P(\hat{y}_n = poc|y = i) \tag{2.2}$$

Thus, $POR_{n,i}$ is the probability of an element of $class_i$ to be classified as the "pass-on-class" poc in the given stage. The values of $R_{n,i}$ and $POR_{n,i}$ depend on the used classifier in the stage and from its operating point. These are both design choices and are hence part of the system optimization. If an input sample is misclassified as the pass-on-class in stage n, this can still be corrected in any of the $n + 1 : N$ stages. If a sample is misclassified MC as any other class in a given stage, the next stages are not powered on and the misclassification leads to a mistake that cannot be corrected. The probability of misclassification of a sample of class i as class j for a given stage $MC_{n,i,j}$ is hence defined as follows:

$$MC_{n,i,j} = P(\hat{y}_n = j \notin \{poc, i\}|y = i) \tag{2.3}$$

This is further illustrated in Fig. 2.2, where for the 3-class second stage the confusion matrix is given as an example. Here, rows indicate the actual class and columns indicate the predicted class. Values classified to the first column of the confusion matrix are passed on to the next classification stage. The confusion matrix further formally illustrates how miss-classified inputs $MC_{n,i,j}$, the recall $R_{n,i}$, and the pass-on-rate $POR_{n,i}$ are defined for each class in this stage. Such a confusion matrix can be generalized to any number of classes I. Note that the rows of the matrix are normalized, such that the matrix contains the classification probabilities of the given true classes instead of absolute counts, simplifying the equations.

The global cost in a hierarchy is a function of several data- and system-level characteristics. First, it is a function of the a-priori probabilities of the occurrence of each class P_i. Second, it also depends on the designed system: the cost C_n of each stage and the pass-on-rates $POR_{n,i}$ of every class in each stage. The average cost per input is hence:

$$C_{1 \to N} = C_1 + \sum_{i=1}^{I} \left(P_i \times \sum_{n=2}^{N} \left(C_n \times \prod_{\eta=1}^{n-1} POR_{\eta,i} \right) \right) \tag{2.4}$$

where $\prod_{\eta=1}^{n-1}(POR_{\eta,i})$ denotes the cumulative product of pass-on-rate for a class i up until stage $n - 1$. In other words, the latter is the fraction of input samples of class i that are passed on to stage n. This factor multiplied with the probability of the class to be present as an input (P_i) and summed over all possible classes (I) is the relative number of times a stage is used per window. Multiplying this with the relative energy cost per stage (C_n) gives the average cost per sample.

The final recall of class i of the full cascade after N stages, $R_{1 \to N,i}$, is independent of a-priori input statistics and can be described as:

$$R_{1 \to N,i} = \sum_{n=1}^{N} \prod_{\eta=1}^{n-1} (POR_{\eta,i}) \times R_{n,i} \tag{2.5}$$

where $R_{n,i}$ is the recall rate for class i in stage n. If a stage n cannot classify samples as class i, i is regarded as a subclass of the pass-on-class p. Thus, $R_{n,i} = 0$ and $POR_{n,i} = POR_{n,poc}$. If recall for every output is considered to be equally important, the average total recall after the end stage is described as:

$$R_{1 \to N} = \text{avg}_i (R_{1 \to N,i}) \tag{2.6}$$

These formulas indicate high recall can either be achieved by high pass-on-rates or by high recall rates. If a sample is passed on to the next stage, it can still be correctly classified by a later stage. However, higher pass-on-rates will also lead to a higher total cost, as this causes more activations of the later stages in the hierarchy.

When the system's average recall from Eq. (2.6) is optimized, the system's precision is also automatically maximized for every class. This is apparent from Eqs. (2.7) and (2.8), which link the amount of true positives tp_i and misclassifications mc_{ij} to total recall $R_{1 \to N}$ and precision $PR_{1 \to N}$. Maximizing total recall and total precision for any general system both require minimizing all final false positives or misclassifications.

$$R_{1 \to N} = \sum_{i=1}^{I} \frac{tp_i}{tp_i + \sum_{j \neq i}^{I} mc_{i,j}} \tag{2.7}$$

$$PR_{1 \to N} = \sum_{i=1}^{I} \frac{tp_i}{tp_i + \sum_{j \neq i}^{I} mc_{j,i}} \tag{2.8}$$

Here, $mc_{i,j}$ is the number of misclassifications, where a sample of class i is misclassified as class $j \neq i$. All of this information can be easily derived from the system's multi-class confusion matrix. This observation is useful, as no formula similar to Eq. (2.5) can be derived for final precision. The above shows that by optimizing for recall in any multi-class system, a high precision is automatically implied.

2.2.3 A Roofline Model for Hierarchical Classifiers

In order to gain insight in the theoretical maximum performance of a classifier in a cascade, we propose a theoretical roofline model. An interesting upper-bound relationship between pass-on-rates and recall can be derived for any classifier in any stage of a hierarchical system. Based on this, we propose a roofline model for hierarchical cascades. Depending on the cost-budget of the classifier in a stage, the relationship between pass-on-rate and recall will be closer or further from the theoretical roofline optimum. In a classifier with a pass-on-class, recall can be described for every class i and for a given stage n as $R_{n,i} = 1 - POR_{n,i} - \sum_{j \neq i}^{I} MC_{n,i,j}$. This formula can be rephrased and be used to plot a relationship between the average recall across all classes except the pass-on-class poc, $\mathrm{avg}_{i \neq poc}(R_{n,i})$, against the average pass-on-rate across all classes, $\mathrm{avg}_i(POR_{n,i})$. It is hence clear that:

$$
\begin{aligned}
&\mathrm{avg}_{i \neq poc}(R_{n,i}) \\
&= 1 - \frac{I}{I-1}\mathrm{avg}_i(POR_{n,i}) + \frac{POR_{n,poc}}{I-1} - \sum_{n,i \neq p}\sum_{j=1}^{I}\frac{MC_{n,i,j}}{I-1}
\end{aligned} \tag{2.9}
$$

In the optimal case, there are no misclassifications. Samples are either passed on or are classified correctly. In this case Eq. (2.9) can be translated into a roofline, as given in Eq. (2.10) and indicated in Fig. 2.3.

$$
\mathrm{avg}_{i \neq poc}(R_{n,i}) = 1 - \frac{I}{I-1}\mathrm{avg}_i(POR_{n,i}) + \frac{1}{I-1}
$$

$$
s.t.\ \mathrm{avg}_i(POR_{n,i}) \geq 1/I \tag{2.10}
$$

Examples of ideal roofline and real curves are given in Fig. 2.3. These real curves are taken from the face recognition hierarchy of Sect. 2.4. The ideal roofline curves are according to Eq. (2.10), while the non-ideal curves are real observed curves, that can

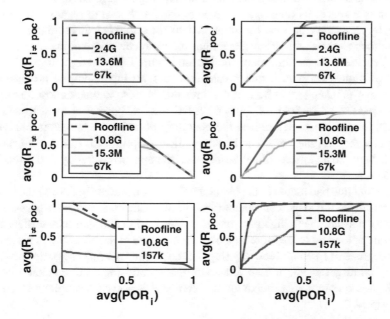

Fig. 2.3 Roofline model and real classifiers for the 2-, 3-, and 10-class classification problems of stages WU, 2 and 3 in Sect. 2.4. The respective cost of every stage in terms of necessary number of operations is given in the legend. More complex tasks require more operations to achieve a performance close to the optimal roofline

be modeled by Eq. (2.9). When the average POR in stage n, $\text{avg}_i(POR_{n,i})$, is below $1/I$, the behavior is non-ideal, as in this operating regime, samples of the pass-on-class are misclassified. The curve reaches a cutting point at $1/I$. Here, all samples are classified correctly (they have an average recall $R_{n,i}$ of 1) and none of them are passed on (they have a pass-on-rate equal to 0) except for samples of the pass-on-class. At an $\text{avg}_i(POR_{n,i})$ larger than $1/I$, the optimal average recall $R_{n,i\neq poc}$ drops, in favor of more passed on samples. This does not lead to catastrophic failures, as any of the next stages can still lead to a correct classification. In the ideal roofline case, the cutting point is only a function of the number of possible classes in a given stage.

In Fig. 2.3, the roofline curves are compared to real performance curves taken from the face classification case study in Sect. 2.4. They are for (a) a binary two-stage wake-up classifier, (b) a 3-class stage, and (c) a 10-class stage. Notice the cutting points at $1/I$ in each roofline. Every full line on these graphs is a classifier with a different complexity and cost. None of these real classifiers is identical to the roofline model, as real classifiers introduce misclassifications. Yet, it is clear that the roofline model offers a very good upper bound, and it is a close match for the most expensive classifier options. Naturally, the higher the complexity and modeling capacity of the classifier, the closer it comes to the optimal roofline model. The deviation of the roofline is mainly due to undesirable misclassifications, which lead to real errors that cannot be compensated for in later stages in the hierarchy.

Every classifier will hence have an associated *operating curve* in the recall vs pass-on-rate space. Its final *operating point* can be chosen to be any point on this curve, by choosing a pass-on-threshold τ. This threshold is a discriminative threshold, similar to the one used in generating binary receiver operating characteristic (ROC) curves. It is the minimum confidence required for a sample to be classified as the pass-on-class *poc*. If this threshold is zero, all input samples are passed on and the $\text{avg}_i(POR_{n,i})$ will be 1. If the threshold is close to one, almost no samples will be passed on in favor of more potential mis- or correct classifications. This threshold value hence determines the operating point of the classifier and can be chosen by the designer, or it can be automatically optimized. In the ideal roofline model, there are no misclassifications. Hence, the threshold directly trades recall for pass-on-rate.

Note that one bad model in the hierarchy does not necessarily destroy the system-level recall, as the stage can operate in a point where it has a high pass-on-rate. However, a system with high pass-on-rates does lead to a higher total cost, as more inputs are passed on to the more expensive later stages in the hierarchy. This leads to many degrees of freedom for every stage in the hierarchy, enabling to optimize cost for a given target recall, or vice versa. Optimally choosing all these characteristics per stage is a complex optimization problem that is solved throughout the rest of this chapter.

Table 2.1 Overview of optimization parameters

Parameter	Comments
C_n	Every cost is associated with a specific classifier in a stage
τ_n	Determines operating point R and POR for a given classifier
N	Number of total stages in the hierarchical cascade

2.2.4   Optimized Hierarchical Cascaded Sensing

Knowing that a maximal recall automatically maximizes precision and based on the previous discussion of general hierarchical sensing systems, a final two-objective optimization problem is defined as follows:

$$\min_{\tau, C} \quad C_{1 \to N}(\tau, C, P)$$

$$\max_{\tau} \quad R_{1 \to N}(C, \tau) \tag{2.11}$$

Here C, P, and τ are vectors representing stage level costs, a-priori class probabilities, and discriminating thresholds. $R_{1 \to N}(C, \tau)$ and $POR_{n,i}(C_n, \tau_n)$ are all determined by the discriminative thresholds τ and the performances of the classifiers (with cost C_n) that can also be chosen. The thresholds τ, C, and the number of stages in the hierarchy N are hence the system's only optimization variables, P is known a-priori. The optimization variables, or hyperparameters, are summarized in Table 2.1.

In Sect. 2.3 this optimization problem is first solved for a general synthetic system, in search for general trends and the influence of all relevant parameters. Section 2.4 is a case study on hierarchical 100-class face recognition.

2.3   General Proof of Concept

In order to prove the proposed hierarchical cascaded processing concept, a generic framework is built. More specifically, for a general system, the impact of input data statistics and system-level specifications on the optimal hierarchy depth and architecture is investigated. This section hence optimizes problem (2.11), using estimated hierarchical POR-R curves in order to derive general trends and hierarchy design recommendations. In Sect. 2.4, the theory is validated on a face recognition case study.

2.3.1 System Description

The used generic system model mimics a cascade of hierarchical classifiers building up to a full 256-class classification system. The cascade is composed of N configurable stages. In each stage, an optimizer will choose an optimal classifier and its operating point. More specifically, in every stage, multiple classifiers are modeled with POR-R trade-off curves such as in the roofline model. The exact operating point is actually determined by the discriminating threshold. The roofline itself is only a theoretical optimum and will hence have an infinite energy cost. In this example, additional non-optimal curves are added that have a lower energy cost, similar to the real curves observed in Fig. 2.3.

The test-setup is a hierarchy that can contain a maximal of $N = 8$ stages, where every stage n classifies 2^n classes, one of which is a pass-on-class. The question to be answered is how many stages the optimal hierarchy contains, and the optimal POR-R trade-off setting for each individual stage. For this setup, 256 different combination of stages are possible in the hierarchy, being all possible architectures with 1–8 stages. The final END stage is fixed as the 256-class classification stage, but the quality of the required classifier in this stage is also flexible.

In order to build a reasonable test case, several assumptions are made. First, costs increase exponentially from stage 1 to 8 according to Eq. (2.12) as the complexity of the classification tasks also increases exponentially.

$$C_n = 10^{log2(I)-1} \tag{2.12}$$

Second, within a single stage, a classifier has an exponentially higher cost if its performance is closer to the theoretical roofline optimum.

This R-POR-C design space is analytically modeled in this example, in order to find a suitable optimum by a steepest descent optimizer. The modeled classifier-to-classifier relative costs are illustrated for stage 1 and $N = 8$ in Fig. 2.4 in the $\text{avg}_{i \neq poc}(R_{n,i \neq poc})/\text{avg}_i(POR_{n,i})$ space, which plot the R-POR trade-off curves for classifiers of different cost. The costs given here take the stage-to-stage cost from Eq. (2.12) into account. Figure 2.4 shows the $\text{avg}_{i \neq poc}(R_{n,i \neq poc})$ vs $\text{avg}_i(POR_{n,i})$, while Fig. 2.4b shows $POR_{poc} = R_{poc}$ vs $\text{avg}_i(POR_{n,i})$ for a binary classifier as a function of cost. The same is shown in Fig. 2.4c, d for the 256-class final stage. Notice none of these stages are ideal: the final average recall is below 1, even for classifiers with a very high cost.

Throughout the rest of this section, three recall–performance cases are discussed. More specifically the section looks into a high-, medium-, and low-recall case, with system-level recalls of 95% (high recall), 85% (medium recall), and 75% (low recall) relative to the theoretical maximum recall of the final END stage.

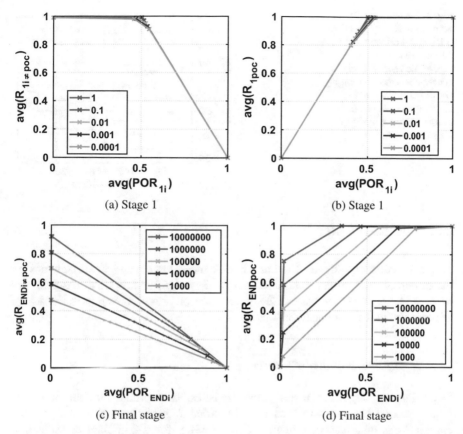

Fig. 2.4 (**a, b**) plot $\mathrm{avg}_{i \neq poc}(R_{1,i \neq u})$ and $R_{1,u}$ versus $\mathrm{avg}_i(POR_{1,i})$ for the first 2-class stage at different costs. (**c, d**) show the same for the 256-class final stage. The legend indicates relative classifier costs

2.3.2   Input Statistics

To estimate the impact of input data statistics, 4 different cases are looked at. A first case is where all input classes are **uniformly** distributed, meaning that they are equally likely to occur. Other cases are either with **medium, highly,** or **extremely** skewed input statistics. The non-normalized probability density functions (PDF) for these different cases are illustrated in Fig. 2.5. Here, some specific classes "C" occur much more frequently than others. An example of such a class "C" can be "noise," "silence" or a common word in speech recognition, or "background" or "owner" in image recognition.

Fig. 2.5 Non-normalized
probability density functions
(PDF) for 4 different cases. A
uniform case where all
classes are equally likely to
appear and a medium, highly,
and extremely skewed
distribution where some
classes are more likely to
appear than others

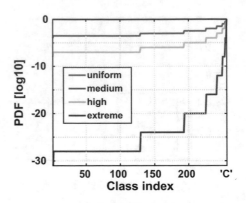

2.3.3   *Experiments*

This subsection investigates the influence of recall targets and input statistics on
the design of an optimal, minimum energy hierarchy. First, optimal cascades are
designed for each of these cases. Second, the section provides insights in the design
of the individual cascading blocks.

2.3.3.1   Optimal Number of Stages

Depending on target recall and input statistics, the optimal number of stages
in a cascaded hierarchy will change. In order to investigate this, all possible
combinations of hierarchies in the $N = 8$ setup for a 256-class classification
problem are optimized for different system targets. More specifically, for every
combination of total recall target, input data statistics assumption, and number of
stages in the hierarchy, a complete classification cascade is optimized using the
steepest descent optimizer. This optimizer selects for every stage in the hierarchy
the optimal R-POR-energy setting, towards minimum system energy. The resulting
energy cost of all chain optimizations is shown in Fig. 2.6. Here, the system-level
energy cost is plotted in function of the selected number of stages in the hierarchy
for different target recall and data statistics assumptions. This shows that for systems
with uniformly distributed input data, shallow hierarchies are optimal. If the input
distribution is skewed more, cost can be reduced by going to deeper hierarchies,
filtering out more samples early on. Costs can vary up to 6 orders of magnitude
depending on input statistics and 2 orders of magnitude depending on the target
recall. The optimal hierarchy architecture varies from 3 to 8 stages. This is further
summarized in Table 2.2, which shows the selected stages in the optimal architecture
with minimal cost. For example, in a system with uniform input distribution and a
high recall target, only the last four stages are used, even though they are the most
expensive. This is because most samples can only get a correct final classification at
the later stages. Thus, early stages with a limited subset of the final output classes

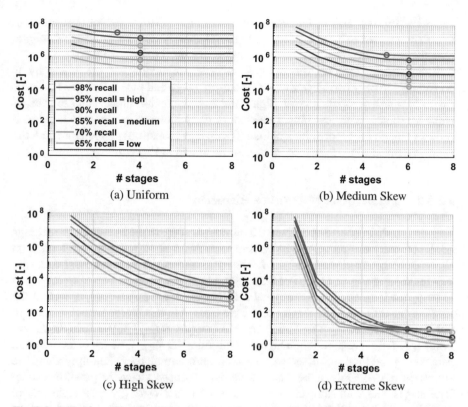

Fig. 2.6 Optimal number of stages in the hierarchy as a function of the relative target recall. The input statistics are varied from (**a**) to (**d**) according to the distributions given in Fig. 2.5

Table 2.2 (x) Stages used in an optimal hierarchy for the 256-class problem. (o) Optimal stage choice if only a wake-up + end stage are allowed

		$S = 1$	2	3	4	5	6	7	END
Rec	Distr.	$I = 2$	4	8	16	32	65	128	256
High	Uni.					x	x	x(o)	x(o)
	Med.			x	x	x	x(o)	x	x(o)
	High	x	x(o)	x	x	x	x	x	x(o)
	Ex.	x(o)	x	x	x	x			x(o)
Med	Uni.					x	x(o)	x	x(o)
	Med.			x	x	x(o)	x	x	x(o)
	High	x	x(o)	x	x	x	x	x	x(o)
	Ex.	x(o)	x	x	x	x	x	x	x(o)
Low	Uni.					x	x(o)	x	x(o)
	Med.			x	x(o)	x	x	x	x(o)
	High	x	x(o)	x	x	x	x	x	x(o)
	Ex.	x(o)	x	x	x	x	x	x	x(o)

are primarily overhead in the case of uniform inputs. However, for skewed input distributions, early stages classify most samples early at a lower costs. This gain outbalances the overhead for the few samples that need to be passed on.

Table 2.2 also shows the optimal cascade architecture if the hierarchy is constrained to only two stages: a first (wake-up) stage and the final stage. The same trends appear here: systems with high skew and low recall targets can use cheaper wake-up stages, with less output classes. A deep hierarchical cascade is up to 3 orders of magnitude more efficient than a two-stage (wake-up) architecture with the same system-level performance, as is also illustrated in Fig. 2.6.

2.3.3.2 Optimal Stage Metrics in a Hierarchy

To gain insight in how to optimize each individual stage in the hierarchy, the settings of all optimally chosen classifier stages in a specific 6-stage medium recall–medium skew hierarchy are plotted in Fig. 2.7. The figure shows the relative cost (1 being the maximum) and $\mathrm{avg}_i(POR_{n,i})$ of the stages, together with their $\mathrm{avg}_{i \neq poc}(R_{n,i \neq poc})$ (1 being the theoretical optimum) and $POR_{n,poc}$ with these settings, as well as the roofline cut-off point $1/I$. Stage 1 and 2 are unused in this hierarchy as indicated in Table 2.2. All operating points are taken at an $\mathrm{avg}_i(POR_{n,i})$ higher than the cut-off point $1/I$. This is required for a high pass-on-rate $POR_{n,poc}$, which is necessary to achieve a good final recall, as these pass-on-rates cumulatively multiply along the stages (Eq. (2.5)). Only in the final stage 8, which doesn't have the possibility to let later stages deal with samples, the system sacrifices $POR_{n,poc}$ for a higher recall $\mathrm{avg}_i(R_{n,i})$. Other optimal hierarchies show similar characteristics across the whole search space.

Fig. 2.7 Relevant design metrics in an optimal 6-stage cascade with medium recall target and high skew

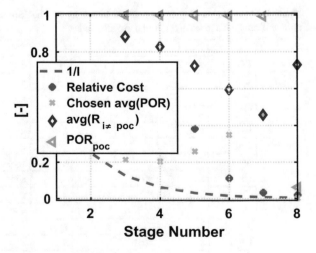

2.3.4   Conclusion

A general hierarchical system classifying 256 output classes was investigated using the developed roofline models and hierarchy optimization framework. This section discussed the influence of the performance goals and of skewed data statistics. With uniformly distributed input data, 4 stages are optimal in our generic model. For increasingly skewed distributions, such as speech or image data, a minimum-cost system will be deeper. This is because easily recognized classes such as "noise" or "background" in that context can be dismissed early in the hierarchy at a low cost using a simple classifier. All of the tested settings using this model require a hierarchy of more than 2-stages. The more conventional wake-up system is hence never optimal.

2.4   Case study: Hierarchical, CNN-Based Face Recognition

2.4.1   A Face Recognition Hierarchy

To illustrate the power of the developed methodology in a real system, we apply it on an actual hierarchical face recognition system using convolutional neural networks (CNN). In the most naive approach, such a system would scan small windows on different scales of a larger input image. Neural network-based large-scale face recognition is very costly (1–2 mJ/250 × 250 window) (Moons et al. 2017b), especially in high resolution images that require a lot of windows to be processed. As input data is generally statistically skewed, it makes sense to build a processing hierarchy to exploit this and reduce the per mean sample cost of the overall system. To illustrate this, regard the pyramid-scale/sliding-window approach on 30 fps Full-HD images (1920 × 1080 pixels) with a window size of 256, a stride of 4, and a scale factor of 2. In 30 fps real-time this approach requires more than 3M window evaluations per second (100k per frame), almost all of which should be classified as backgrounds. At 1 mJ/window evaluation, this consumes 3 kW's of power, which is obviously infeasible on a wearable, battery-constrained device. As distinguishing faces from backgrounds is a much simpler task that can be performed at ≈1 uJ/subsampled 32 × 32 window (Bankman et al. 2018), such a detector can be used as a wake-up stage for the more complex and costly face recognizer. Only if the face-detector detects a face, the more costly subsequent face recognizers are used. If not, the system goes on to the next window, reducing the average cost per window considerably depending on the input data's statistics. While it is clear that a hierarchical system can bring significant benefits for this face recognition system, it is not clear how many stages should be used in the hierarchy. If there is still statistical skewness between different face classes, e.g., the owner of a device appearing much more frequently than other faces, intermediate stages could be added to exploit this

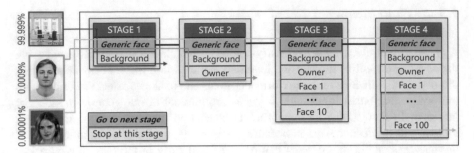

Fig. 2.8 Illustration of the tested hierarchical face recognition example. Backgrounds and owners are much more likely to appear than specific faces, but also easier to detect with cheaper classifiers

skewness in an effort to further decrease the activation rate of the costly final stage. On top of that, the R-POR-energy settings of each stage should be tuned optimally for minimal system energy consumption.

To study this, Fig. 2.8 illustrates the generalization of this wake-up approach to a hierarchical N-stage system that can ultimately distinguish between 100 faces. A window is passed on to the next stage of the hierarchy, only if it is classified as the "pass-on-class." In all other cases, i.e., when the image is classified as a "*background*," the "*owner*," or another specific "*$face_i$*" the window is considered to be classified. Again here, as the tasks in stage $n <$ END are cheaper than the ultimate task END and the distribution is skewed, cost can be reduced significantly. In this case study, the general framework developed in Sect. 2.2 is used to optimize the full hierarchy towards minimal average cost per sample, while maintaining overall face recognition accuracy. This analysis proves expanding the typical two-stage wake-up system to a multi-stage hierarchical cascade reduces cost considerably. Optimal operating points for every stage are also derived.

In order to analyze this specific case, the 100 face recognition system is split into a maximum of $N = 4$ stages (face detection WU—owner detection S2—10 faces S3—100 faces END), where the last stage END is the full 100 face recognizer. For each of those stages, 15 CNN-based parametrized models are trained with varying complexity: the subsampled dimensions of the input image, the number of layers per network (the network depth), and the number of filters per layer (the network width). Larger networks operating on larger input images generally achieve higher recall and precision on the input data set, albeit at a higher cost which is regarded as computing complexity in this case. The general architecture and parameters of these networks are given in Table 2.3, similar to the parameters used in Moons et al. (2017a). In these networks, the input dimensions, the total number of layers (total depth), and the width of those layers are varied. The smallest network takes a subsampled 32×32 RGB input and classifies it using a 4-layer CNN with 4 filters per layer. The largest network takes a 128×128 RGB input and classifies it using a 7-layer network with up to 1024 filters per layer. All networks are trained

Table 2.3 Parametrized network topology used for all N stages in hierarchical face recognition

Block	(W,H) Dim.	Kernel	Stride	# Layers	Width
In. Layer	32–64–128	3–5–7	1–2–4	1	4–256
Block A	32	3	1	1	4–256
MaxPool	32	2	1	1	–
Block B	16	3	1	1–2	4–512
MaxPool	16	2	1	1	–
Block C	8	3	1	1–3	4–1024
MaxPool	8	2	1	1	–
Dense	$16 \times width_C$	–	–	1	–

All intermediate activations are LeakyRelu, the dense layer uses a softmax activation function

on a 100-face subset of the VGG FACE-2 data set (Cao et al. 2017). We use batch normalization and random data augmentation (shear, channel shifts, width shifts, height shifts, zoom, and horizontal flips) to prevent overfitting.

2.4.2 Hierarchical Cost, Precision, and Recall

Each individual trained convolutional neural network model has a specific recall and pass-on-rate for each of its output classes, together with an associated cost. All these terms are defined as in Sect. 2.2.

According to the framework discussed in Sect. 2.2, we minimize full system cost, Eq. (2.4), while maximizing total system recall as in Eq. (2.6). As explained in Sect. 2.2.2, precision will be automatically optimized as well.

As shown by Eq. (2.5), the full recall per class R_i depends on recall in every stage $R_{n,i}$ and the cumulative product of $POR_{n,i}$ in the previous stages. Note that theoretically the values of neither $POR_{n,i}$ and $R_{n,i}$ can be taken independently of the previous stages, as they influence the images that are offered to subsequent stages. Difficult images will be passed on more often than easy images, as they will be easily classified by one of the cheap first stages. However, in this optimization we assume the POR and R of every stage to be independent of the previous stages in order to have an analytical closed formula. We then later experimentally verify if the results under these assumption are correct, which is true in this test case. Hence, the shown values for $POR_{n,i}$ and $R_{n,i}$ in this section are based on a representative test set in the full hierarchy and not only based on the performance of the individual stages.

The performance of some individual classifiers in the different stages can be visualized using the proposed roofline curves, as discussed in Sect. 2.2.3 and illustrated in Fig. 2.3. Here, the performance of the classifying stage, embedded in the hierarchy, is given in the POR-vs-$Recall$ space. It can, e.g., be seen that, for the third stage, the best classifier (blue) requires $69k\times$ more operations than the cheapest depicted classifier, but at a performance that is much closer to the ideal roofline.

In order to estimate the total cost, realistic a-priori input statistics have to be derived for all classes. Here, it is assumed that the system under investigation is used in the ESAT/MICAS—KU Leuven office in a surveillance context. In this office, on average 100 persons pass through the corridor every hour. Each of those people is in view of the system for 5 s. The a-priori probabilities for each subgroup in this scenario are taken as [1 14.000.000 100 10 1]/(14.000.112) = ["pass-on-class" (F), "background" (BG), "owner" (O), "face2-10," "face11-100"]. This is representative if the system investigates FHD images at 30 fps with window sizes of 256, a stride of 16, and a pyramid-scale factor of 2. When processing these windows, they are down sampled to one of the supported window sizes given in Table 2.3: either 32×32, 64×64, or 128×128 RGB.

2.4.3   An Optimized Face Recognition Hierarchy

Once all relations between the overall cost $C_{1 \to N}$, pass-on-rates $POR_{n,i}$, and global recall R_i are found, the hierarchy can be optimized towards minimum cost (in terms of average amount of operations per sample) and maximum overall recall. The variables in this optimization problem are the chosen classifier and the "pass-on-class" detection thresholds at each stage, as these two factors determine $R_{n,i}$ and $POR_{n,i}$. A particle swarm multi-objective optimizer is used to numerically solve this discrete optimization problem. Figure 2.9 shows the results of this optimization for a 4-stage face recognition pipeline.

Figure 2.9a shows all pareto-optimal possible systems in the efficiency-vs-recall and efficiency-vs-precision design space. Here, efficiency is described as [Sample/op] or the number of samples that can be classified per MAC-operation. For the same recall, the 4-stage system obviously requires orders of magnitude less operations than a system existing out of the END stage only. For example, its efficiency at an average recall of 80% is 10.000 times higher than in the single stage case. Figure 2.9a also experimentally verifies that optimizing for high average recall also automatically optimizes for high average precision, a claim made in Sect. 2.2. If recall is high and close to its maximum, precision is also maximized. Precision and recall are high in the 4-stage system over a wide range of efficiencies, while they are only high at a very low efficiency in the 1-stage system.

The characteristics for every pareto-optimal architecture in Fig. 2.9a are illustrated in Fig. 2.9b. Figure 2.9b shows the optimal chosen classifiers, the pass-on-rate for every class at every stage ($POR_{n,i}$), and the final recall per class, averaged per subgroup with equal a-priori probability for readability. The figure clearly indicates that systems with higher recall also require more expensive building blocks. Even at high recall, the complexity of the wake-up stage is 3 orders of magnitude lower than the complexity of the END stage, which explains most of the gains of the multi-stage system. All used classifiers operate near the roofline.

We expand this analysis by performing the same optimization on shallower hierarchies, from 1 stage to the maximum of 4 stages. Figure 2.10 illustrates the

Fig. 2.9 A comparison of hierarchical 100-face recognition using different stages in the recall vs efficiency [Frames/op] space. The upper right corner is optimal. (**a**) Recall and precision versus efficiency. (**b**) Complexity, POR and system recall for every system in the optimization

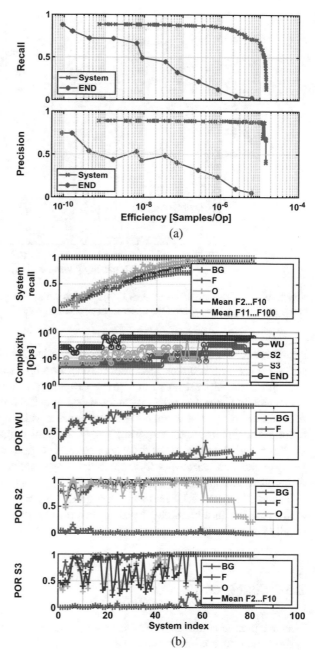

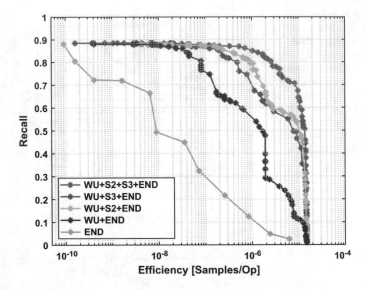

Fig. 2.10 A comparison of hierarchical 100-face recognition using different stages in the recall vs efficiency [Samples/op] space. The upper right corner is optimal

performance of these different hierarchical architectures in the efficiency-vs-recall space for the full 100 face recognition functionality. An architecture is optimal if it achieves the highest recall at a given efficiency in terms of processed samples per operation (Samples/op). Obviously, using only the END stage is very inefficient. Also the more typical 2-stage hierarchy with a wake-up stage proves to be suboptimal for this test case, requiring 1-to-2 order of magnitude more operations at similar recall than the optimal 4-stage case. The performance gains achieved by going to deeper hierarchies with more than 4 stages are arguably minimal, especially if the maximum latency is taken into account.

Figures 2.9a and 2.10 also show the quality-vs-energy trade-off can be dynamically changed in a hierarchy, by changing the threshold that determines POR and recall of every stage.

2.5   Conclusion

This chapter generalizes wake-up systems to multi-stage hierarchical cascaded systems that have a lower system-level cost and are better adapted to skewed data with non-uniform probability distributions. It is an approach that minimizes the energy consumption of a neural network-based detection system on the **application level**, rather than on the algorithm (see Chap. 3) or hardware level (see Chaps. 4, 5, and 6).

This chapter offers four key contributions.

1. Wake-up-based recognition systems are generalized to multi-stage hierarchical cascades.
2. A design framework is presented that can be used to simultaneously optimize performance and minimize costs in hierarchical cascades. Linked to this is a theoretical roofline model providing insight in the performance of the individual stages in the hierarchy.
3. General trends in hierarchical cascades are derived through analyzing a generic example. It is shown that, while hierarchical cascades do not bring significant benefits for uniform input data statistics, systems with skewed input data statistics, such as speech and object detection tasks, benefit from deeper cascades. If an intermediate stage is used, its optimal operating point is close to the theoretical roofline. Stages with bad recall and pass-on-rate performance are never beneficial.
4. The approach is further illustrated by designing a 4-stage 100-face recognition application. An optimal operating point exists, where 4 orders of magnitude in cost-efficiency can be gained compared to the single-stage classifier and 2 orders of magnitude compared to the traditional 2-stage wake-up-based system.

The proposed framework and roofline model are generally applicable on many sensory data applications. I hope this framework can contribute in moving this field from ad-hoc designs towards automated system optimizations.

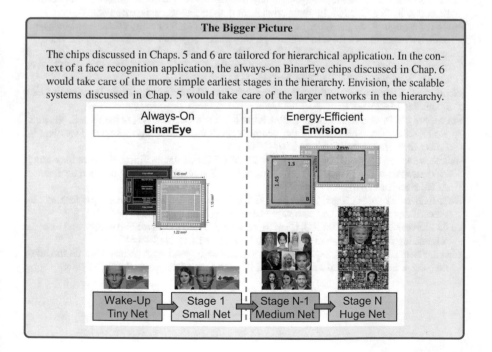

The Bigger Picture

The chips discussed in Chaps. 5 and 6 are tailored for hierarchical application. In the context of a face recognition application, the always-on BinarEye chips discussed in Chap. 6 would take care of the more simple earliest stages in the hierarchy. Envision, the scalable systems discussed in Chap. 5 would take care of the larger networks in the hierarchy.

Always-On
BinarEye

Energy-Efficient
Envision

Wake-Up
Tiny Net

Stage 1
Small Net

Stage N-1
Medium Net

Stage N
Huge Net

References

Badami K, Lauwereins S, Meert W, Verhelst M (2015) Context-aware hierarchical information-sensing in a 6uw 90nm CMOS voice activity detector. In: 2015 IEEE international solid-state circuits conference-(ISSCC). IEEE

Bankman D, Yang L, Moons B, Verhelst M, Murmann B (2018) An always-on 3.8uj/classification 86accelerator with all memory on chip in 28nm cmos. In: ISSCC technical digest

Cao Q, Shen L, Xie W, Parkhi OM, Zisserman A (2017) Vggface2: a dataset for recognising faces across pose and age. ArXiv preprint arXiv:171008092

Ghahramani Z, Jordan MI, Adams RP (2010) Tree-structured stick breaking for hierarchical data. In: Advances in neural information processing systems, pp 19–27

Goetschalckx K, Moons B, Lauwereins S, Andraud M, Verhelst M (2018) Optimized hierarchical cascaded processing. IEEE J Emerging Sel Top Circuits Syst. https://doi.org/10.1109/JETCAS.2018.2839347

Han S, Mao H, Dally WJ (2015) Deep compression: compressing deep neural networks with pruning, trained quantization and Huffman coding. ArXiv preprint:151000149

Huang G, Che D, Li T, Wu F, van der Maaten L, Weinberger K (2017) Multi-scale dense networks for resource efficient image classification. ArXiv preprint arXiv:170309844, submitted to ICLR 2018

Li M, Bijker W, Stein A (2015) Use of binary partition tree and energy minimization for object-based classification of urban land cover. ISPRS J Photogramm Remote Sens 102:48–61

Liu L, Deng J (2017) Dynamic deep neural networks: optimizing accuracy-efficiency trade-offs by selective execution. ArXiv preprint arXiv:170100299

Moons B, Goetschalckx K, Van Berckelaer N, Verhelst M (2017a) Minimum energy quantized neural networks. In: Asilomar conference on signals, systems and computers

Moons B, Uytterhoeven R, Dehaene M Wim Verhelst (2017b) Envision: a 0.26-to-10 tops/w subword-parallel dynamic-voltage-accuracy-frequency-scalable convolutional neural network processor in 28nm FDSOI. In: International solid-state circuits conference (ISSCC)

Panda P, Sengupta A, Roy K (2015) Conditional deep learning for energy-efficient and enhanced pattern recognition. CoRR abs/1509.08971, http://arxiv.org/abs/1509.08971, 1509.08971

Price M, Glass J, Chandrakasan AP (2017) A scalable speech recognizer with deep-neural-network acoustic models and voice-activated power gating. In: 2017 IEEE international solid-state circuits conference (ISSCC). IEEE, pp 244–245

Saberian M, Vasconcelos N (2014) Boosting algorithms for detector cascade learning. J Mach Learn Res 15:2569–2605

Sun M, Snyder D, Gao Y, Nagaraja V, Rodehorst M, Panchapagesan NS, Matsoukas S, Vitalade-vuni S (2017) Compressed time delay neural network for small-footprint keyword spotting. In: Proceedings Interspeech 2017, pp 3607–3611

Venkataramani S, Raghunathan A, Liu J, Shoaib M (2015) Scalable-effort classifiers for energy-efficient machine learning. In: Proceedings of the 52nd annual design automation conference. ACM, New York, p 67

Viola P, Jones M (2001) Rapid object detection using a boosted cascade of simple features. In: Proceedings of conference on computer vision and pattern recognition, pp 511–518

Xu ZE, Kusner MJ, Weinberger KQ, Chen M, Chapelle O (2014) Classifier cascades and trees for minimizing feature evaluation cost. J Mach Learn Res 15(1):2113–2144

Yuan J, Chan HY, Fung SW, Liu B (2009) An activity-triggered 95.3 db dr −75.6 db thd cmos imaging sensor with digital calibration. IEEE J Solid State Circuits 44(10):2834–2843

Chapter 3
Hardware-Algorithm Co-optimizations

This chapter is structured as follows. Section 3.1 gives a non-exhaustive overview of SotA techniques used to optimize hardware platforms for deep learning algorithms: (A.) dataflow optimizations, (B.) sparse optimizations, and (C.) exploitation of a neural network's fault-tolerance. The rest of the chapter focuses on the system-level aspects of (C.). Section 3.2 is a high-level energy model to estimate the influence of fixed-point computations on the energy consumption of neural networks. This model is based on the circuit-level techniques developed in Chap. 4 and the measurement results on the chips designed in Chap. 5. It is used here in an analysis on test-time fixed-point neural networks in Sect. 3.3 and train-time quantized neural networks in Sect. 3.4. The latter networks are used to design minimum energy networks, by simultaneously optimizing the network topology and the hardware architecture it would run on. Finally, Sect. 3.5 uses the same model to estimate the effect of nonlinear quantization techniques.

3.1 An Introduction to Hardware-Algorithm Co-optimization

This section is a non-exhaustive literature review of relevant techniques on the algorithmic and architectural level used to minimize the energy consumption in neural networks. All of the discussed techniques and implementations exploit one of the three key neural network characteristics (A,B,C):

A. Deep learning networks exhibit a very particular dataflow with a large amount of potential parallelism and data reuse. This can be heavily exploited in application specific hardware designs.

© Springer Nature Switzerland AG 2019
B. Moons et al., *Embedded Deep Learning*,
https://doi.org/10.1007/978-3-319-99223-5_3

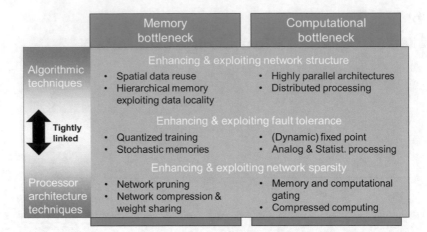

Fig. 3.1 An overview of the algorithmic and processor architecture techniques discussed to increase efficiency and enable the inference of deep neural networks in embedded devices

B. Deep learning networks demonstrate large sparsity: Many parameters become very small, or even equal to zero after network training. Also many data values propagated with the network during evaluation become zero. This can be exploited to reduce operations and memory fetches in hardware, yet can also be stimulated further with innovative training techniques.
C. Deep learning networks prove to be robust to approximations or fault introductions. This is exploited in various reduced precision hardware implementations as discussed above. Ultimately this observation leads to very specific dynamic precision hardware architectures tailored to deep learning in Chaps. 4 and 5, as well as to specific analog circuits and architectures for low-power neural networks in Chap. 6.

Some examples here are pure hardware optimizations made to run given neural networks more efficiently. Other works focus on redesigning or retraining slightly changed network architectures that perform better on a new hardware platform. An overview of the highlighted techniques is given in Fig. 3.1.

3.1.1   Exploiting Network Structure

In many application areas, designers have improved energy-efficiency of algorithm evaluation by moving away from general purpose processors, and developing customized hardware accelerators, specifically for the target algorithms. Such accelerators can exploit the known dataflows within the algorithm, to (1) enhance the parallel execution of the algorithm, as well as (2) minimize the data movement Fig. 3.2. Several ASICs targeting the efficient execution of convolutional and

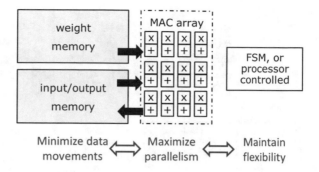

Fig. 3.2 Custom deep neural network processors gain efficiency by minimizing data movements and maximizing parallelism. Still, it is crucial not to lose all flexibility in mapping a wide variety of networks. FSM stands for finite-state-machine

fully connected layers have recently been published. All solutions exhibit a very large degree of parallelization, far beyond CPU parallelism. This demonstrates itself in a datapath easily containing a few hundreds to thousands multiply accumulators (MAC), with Google's recent tensor processing unit (TPU) as an extreme example with 64,000 MACs (Jouppi et al. 2017).

Providing data to all these functional units in parallel would be nearly impossible, if the temporal and spatial locality of the data was not exploited. Indeed, many computations within one network layer share common inputs. More specifically, as highlighted in Fig. 1.5's pseudo-code for a "plain" CNN, every weight parameter is reused approximately M^2 times across multiple convolutions of the same slice in the output tensor, and every input data point is reused across F different slices of the output tensor. Moreover, the intermediate accumulation results o have to be accumulated $C \times k^2$ times. This can be exploited in several ways in a custom accelerator to further boost efficiencies beyond the highly parallel, yet not dataflow optimized GPU's.

On one hand, data reuse can be exploited by reusing the same data across multiple parallel execution units or, equivalently, across multiple time steps on the same execution unit. In this topology, three extreme cases can be distinguished.

- In the "weight parallel" or **"input stationary"** approach (Fig. 3.3) the same input data is multiplied with several weights of different output channels of a layer. Here, ideally every input will only be loaded once into the system. This however has negative repercussions on the weight memory bandwidth (BW), as the weights have to be reloaded frequently every time a new input is applied. Moreover, the accumulation of the output cannot be done across different clock cycles, requiring intermediate results to be pushed into memory, and re-fetched later, strongly impacting the input/output memory bandwidth.
- The **"weight stationary"** or "input parallelism" approach improves the weight memory bandwidth, yet at the expense of the input memory bandwidth. Here, every weight is fetched once and multiplied with many input values.
- The **"output stationary"** scheme reloads new weights and inputs every single clock cycle, yet is able to accumulate the intermediate results across different clock cycles, to the benefit of the output memory bandwidth.

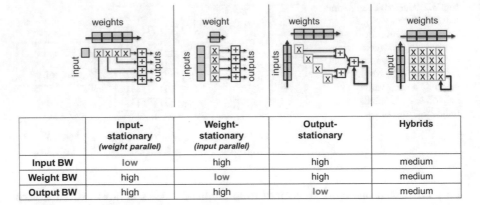

	Input- stationary (weight parallel)	Weight- stationary (input parallel)	Output- stationary	Hybrids
Input BW	low	high	high	medium
Weight BW	high	low	high	medium
Output BW	high	high	low	medium

Fig. 3.3 Different computer architectures allow data reuse to be maximized, reusing either inputs, weights, intermediate results, or a combination of the three

In practice most realizations implement a hybrid form of the three presented extreme cases. Moons and Verhelst (2016), an example discussed in Chap. 5, is a hybrid input–output and weight stationary implementation. In Shin et al. (2017) input and output are stationary to minimize the chip IO bandwidth. In Chen et al. (2016a) and Peemen et al. (2013) a more elaborate overview of the different parallelization schemes can be found, together with an assessment of their merits. Other notable early implementations are Conti and Benini (2015), Farabet et al. (2011), and Chen et al. (2014). All of these examples are optimized specifically for CNN dataflows.

A complementary way to reduce the energy burden of continuous data fetches is to reduce the energy cost of every data fetch exploiting temporal data locality. Most realistic deep networks require so much weight and input/output memory (Mbytes to Gbytes), that it is impossible to fit them in on-chip memory, hence requiring fetches from energy-costly external DRAM. Similar to traditional processors, this can however be mitigated by a memory hierarchy of one or more levels of on-chip SRAM or register files. Frequently accessed data can as such be stored locally to reduce its fetching cost (Fig. 3.4).

An important difference with general purpose solutions is however that the sizes of the memories in the hierarchy can be optimized towards the network's structure, e.g., foreseeing local memory capable of caching exactly one weight tensor. More importantly even, the networks can be trained with the processor memory hierarchy in mind. As such, networks have, e.g., been explicitly trained to completely fit in on-chip memory. This optimization is of course highly interwoven with the parallelization scheme. By jointly optimizing them, one can adjust the degree of parallelization to the memory hierarchy to minimize the product of both the number of memory accesses with the cost of every memory access (Cecconi 2017). Distributed and **systolic** processing can be seen as an extreme of such hierarchical memories. In the systolic processing concept, a two-dimensional array of functional units processes data locally and passes inputs and intermediate results from unit to

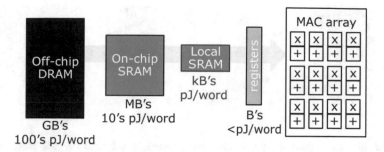

Fig. 3.4 A well-designed memory hierarchy avoids drawing all weights and input data from the costly DRAM interface and stores frequently accessed data locally

unit, instead of to/from global memory. These functional units are each equipped with a very small SRAM (as in Chen et al. 2016b), or even only registers (as in Jouppi et al. 2017), to store data locally and maximize data reuse within the array. Processing happens as a systolic wavefront through the array, where weight coefficients can be kept stationary in the functional units, input data is shifted in one direction through the array, and output data accumulates in the orthogonal direction. This allows performing a very large number of computations for convolution or matrix multiplication in parallel, by keeping all systolic elements busy without burdening the memory bandwidth. Interested readers are pointed to Annaratone (1987) and Jouppi et al. (2017) for more details.

Another approach exploiting locality is **in-memory computing**, where the computation is integrated inside a memory array (Biswas and Chandrakasan 2018; Gonugondla et al. 2018). While this is also pursued in traditional memory architectures, results look especially promising for emerging non-volatile memory arrays. For example, in resistive memory technologies (Shafiee et al. 2016) or Flash memories. An example designed by this author (among others) is given in detail in Chap. 6. The latter is an example of hardware-algorithm co-optimization, as several of the techniques used to make that design efficient require a hardware-specific neural network topology and training algorithm.

3.1.2   Enhancing and Exploiting Sparsity

A second opportunity for hardware platforms to be adapted to neural network algorithms is by exploiting their high levels of sparsity. Many of the weight values, as well as intermediate data values, are zero. Figure 3.5a shows the sparsity of AlexNet. As can be seen more than 70% of the activations are zero. At reduced bit width computations, also several weight values are quantized to zero. This opens up many opportunities: On the hardware side, this can be exploited by (Fig. 3.5b) (1) preventing any MAC with a zero-valued input, as this multiply and accumulate with zero would only waste energy; (2) not even fetch zero-valued data values from

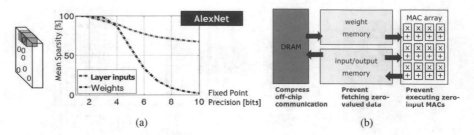

Fig. 3.5 (**a**) The sparsity of inputs and weight values of a typical network in function of computational precision at which the network is evaluated. (**b**) This sparsity allows energy to be saved in the processor's input/output interface, on-chip memories, and datapath

memory; and (3) strongly compress the on/off chip datastream using, e.g., Huffman or other types of encoding. Several hardware implementations exploit these CNN-characteristics, among which the ones discussed in Chap. 5. Moons and Verhelst (2016) and Chen et al. (2016b) skip all unnecessary sparse operations by gating the inputs to their arithmetic units if the input data is zero and compresses off-chip datastreams. Commercial products apply similar techniques. The architectures presented in Albericio et al. (2016) and Kim et al. (2017) allow speeding up sparse network evaluation.

Deep network training algorithms can also be modified to enhance the networks sparsity, by iteratively pruning the smallest weight values (quantizing them to zero), and retraining the network (Han et al. 2015b). One step further, energy-aware pruning techniques even take the energy consumption model of the hardware into account, to maximize pruning efficacy (Sze et al. 2017). Traditional accelerators can benefit from such compression, but only in terms of a reduction in memory size and amount of memory accessed. The EIE engine however (Han et al. 2016a) demonstrates that it is also possible to operate directly on the compressed data by adapting the datapath and memory interface to the compressed data format. In Yin et al. (2017), the deep-compression concept is expanded upon by enforcing neural networks to be structurally sparse, instead of randomly sparse in Han et al. (2015a). Other techniques are based on singular value decomposition (SVD) (Xue et al. 2013) or a combination of SVD and deep compression (Goetschalckx et al. 2018).

If a hardware platform that can exploit sparsity is available, networks can be designed to be as sparse as possible. One approach is to use ReLU activation functions and to force weights to be zero or small through L1 or L2 regularization, respectively (see Sect. 1.3.4 in Chap. 1).

3.1.3 Enhancing and Exploiting Fault-Tolerance

A third important aspect of deep neural networks which can be exploited in custom processor designs is their fault-tolerance. Many studies observe the robustness of

CNNs and other networks to perturbations on their weight parameters and their intermediate computational results (Moons et al. 2016; Gysel et al. 2016). This can be exploited both to develop more efficient hardware (Chaps. 4, 5 and 6). On the other hand, neural networks can also be explicitly trained to perform better on such platforms, as is illustrated in Sect. 3.4 in this chapter.

A straightforward way to benefit from the network's fault-tolerance is to perform the computations at reduced computational accuracy with limited recognition loss. Typical benchmarks can be run at 1–9 bit fixed-point rather than 32b floating-point at less than 1% accuracy loss (Moons et al. 2016; Jiang and Gielen 2003; Dundar and Rose 1995). This is possible by quantizing all weights of a floating-point trained network before execution, as is discussed in Sect. 3.3. Improved results can be obtained when already introducing quantization during the training step itself (Hubara et al. 2016b; Moons et al. 2017a), resulting in smaller or lower precision networks for the same application accuracy (Sect. 3.4). If the network architecture and hardware platform are jointly optimized, as is the case in Sect. 3.4, it is a hardware-algorithm co-optimization. As such, these techniques have been a major part of my work through this PhD. The algorithmic side of this is discussed in detail in this chapter. The possible energy gains in hardware are discussed in detail in Chaps. 4 and 5

At its most extreme, networks have been specifically trained to operate with only 1b representations of weights only (Rastegari et al. 2016) and both weights and activations (Rastegari et al. 2016; Courbariaux and Bengio 2016) where all multiplications can be replaced by efficient XNOR operations (Andri et al. 2016, 2017). In Rastegari et al. (2016), a binary-weight version of ImageNet is only 2.9% less accurate (in top-1 measure) than the full-precision AlexNet (Krizhevsky et al. 2012a). Chapter 6 discusses the BinaryNet ASIC implementations I've worked on as part of my PhD.

As a summary and introduction to Chap. 4, this observation of fault-tolerance can lead to major energy savings, compared to the current CPU and GPU architectures that operate using 32-16b floating-point number formats. Reducing precision from 32b floating-point to lower fixed-point precision does not only reduce computational energy, but also minimizes the storage and data fetching cost needed for network weights and intermediate results. Moreover, for very low bit widths, this even allows to replace multipliers that have several data values with a common weight factor, by preloaded look-up tables (Shin et al. 2017). While most processors operate at constant 16-, 12-, or 8-bit word lengths, some recent implementations support variable world length computations, where the processor can change the used computational precision from operation to operation (Moons and Verhelst 2016; Shin et al. 2017; Moons et al. 2017c; Lee et al. 2018). This accommodates for the observation that the optimal word length for a deep network strongly varies from application to application, and even differs across various layers of a single deep network (Moons et al. 2016) (Fig. 3.6a). The variable-resolution processors use a technique termed dynamic-voltage-accuracy-frequency-scaling (DVAFS) (Moons and Verhelst 2015; Moons et al. 2017b) to jointly reduce the switching activity, the supply voltage, and the parallelization scheme when computational resolution

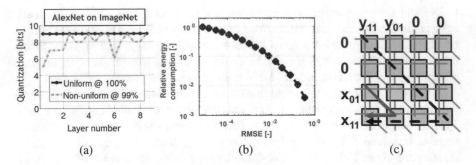

(a)   (b)   (c)

Fig. 3.6 (**a**) When quantizing all weight and data values in a floating-point AlexNet uniformly, the network can run at 9-b precision. Lower precision can be achieved without significant classification accuracy loss by running every layer at its own optimal precision. This allows (**b**) saving power in function of computational precision and (**c**) building multipliers whose energy consumption scales drastically with computational precision, through reduced activity factors and critical path scaling. This is further discussed in Chaps. 4 and 5

drops (Fig. 3.6b). This results in a scaling of the systems energy consumption which is super-linearly with the computational resolution (Fig. 3.6c). The technique is discussed in detail in Chaps. 4 and 5.

Another school of thoughts targets energy savings through tolerating non-deterministic, statistical errors. This can be done by executing the convolutional kernels in the noisy analog domain (Fick et al. 2017). Alternatively, in the digital domain, stochastic fault-tolerance can be exploited by operating the circuits (Lin et al. 2016; Bankman et al. 2018) and/or memory (Whatmough et al. 2017; Yang et al. 2018) in the energy-efficient near threshold regions.

Finally, also the operational circumstances can strongly influence the networks tolerance to approximations. In a given classification application, the quality of the inputs might change dynamically, or some classes might be easier to observe than others. If one tries to learn one common network, which performs acceptable under all possible circumstances and classes, a large, complex, energy hungry network topology would be needed. Recent work however promotes to train hierarchical or staged networks (Huang et al. 2017), which perform classifications in several optional stages. At each stage, only a few layers of the network are executed, after which a classification layer tries to guess the class from the current outputs. Only if the obtained probabilities are not outspoken enough, additional network layers and classifiers are run, until a classification with distinct probabilities is obtained. Such dynamic evaluations can be performed on any hardware platform, but again heavily benefits from implementation-aware training techniques, or topology-optimized implementations. In inference on the IMAGENET data set, (Huang et al. 2017) need up to 2.6× less operations than the prior SotA networks at iso-accuracy. My own approach to such hierarchical processing is discussed in Chap. 2.

The rest of this chapter focuses on the system-level aspects—accuracy and high-level energy costs—of different types of reduced precision neural networks: networks operating at arbitrarily low fixed-point precision. Chapter 4 illustrates the energy gains that can be expected in QNNs in real hardware. Chapter 5 discusses two chip implementations that benefit from the technique. Finally, Chap. 6 discusses several physical implementations of the most extreme form of QNN: binary (1b) neural networks.

3.2   Energy Gains in Low-Precision Neural Networks

Exploiting a neural network's fault-tolerance through precision scaling lowers the active power consumption of a digital circuit. Hardware and circuit-level considerations required for applying this in a real system are discussed in Chaps. 4 and 5. However, a basic energy model is useful here already.

The total power consumption in a digital system consists of a dynamic and a leakage component. It can be summarized as $P = \alpha C f V^2 + P_{\text{leakage}}$ where α is the circuit switching activity, f is the clock frequency, C is the total switching capacitance, and V is the circuit's supply voltage. By scaling precision (i.e., dynamically scaling the number of bits that encode the network's weights and inputs), the switching activity α can be substantially reduced. Figure 3.6b illustrates the achievable energy-accuracy trade-off in a typical digital multiplier. Note that a striking 12× energy gain can be achieved at an output error of 1% root-mean-square-error (RMSE). As neural networks prove to be fault-tolerant, their performance may not degrade under these RMSE deviations in the intermediate computations. In order to estimate the energy consumption of reduced precision networks more accurately, a generic hardware energy model based on Chap. 5 is built below and in Fig. 3.7.

Global energy per inference is the sum of the energy consumed in communication with an off-chip DRAM and the energy consumption of the processing platform itself. The total energy consumed per network inference is then:

$$E_{\text{inf}} = E_{\text{DRAM}} + E_{HW} \tag{3.1}$$

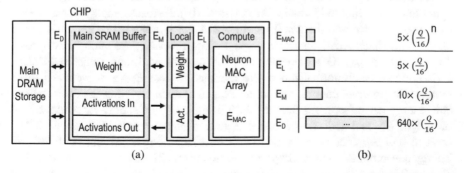

(a)   (b)

Fig. 3.7 Used energy model for a generic hardware platform. (**a**) High-level overview of the system architecture. (**b**) Relative energy consumption per equivalent multiply–accumulate (MAC) operation (E_{MAC}), read/write from the local (E_L) and main (E_M) SRAM buffers and per read/write from a large DRAM memory (E_D) of an intQ word

3.2.1   Energy Consumption of Off-Chip Memory-Access

The available memory in an always-on chip is inherently limited due to costs and
leakage energy constraints and hence typically insufficient to store full models **and**
feature maps. If this is the case, the chip will constantly have to communicate with a
larger off-chip memory system. The cost of this interface is two orders of magnitude
higher than a single equivalent MAC-operation (Horowitz 2014). Using less bits
for weights and activations can hence be potentially more energy-efficient, if the
achieved network compression, both for weights and activations, makes the network
fit completely in on-chip memory. Off-chip DRAM access energy is modeled as:

$$E_{\text{DRAM}} = E_D \times (s_{in}^2 \times c_{in} \times M/Q + 2 \times f_{re-fetch} + w_{re-fetch}) \qquad (3.2)$$

where E_D is the energy consumed per intQ DRAM access, as in Fig. 3.7. s_{in},
c_{in}, and M/Q are respectively the input image's dimensions, the number of input
channels, and the first-layer factor defined in Sect. 3.4.1.2. $f_{re-fetch}$ and $w_{re-fetch}$
are the number of words that have to be re-fetched/stored from/to DRAM if an
intermediate feature map or model does not fit on-chip. These factors depend on the
used tiling scheme and the available amount of on-chip memory.

3.2.2   Generic Hardware Platform Modeling

The hardware platform, shown in Fig. 3.7, is a typical processing platform for
CNNs. It contains a parallel neuron array, with a fixed area for p MAC units and
two levels of on-chip SRAM memory. A large main buffer enables storing M_W-
bits of weights and $M_A = M_W$-bits of activations, of which 50% is used for the
current layer's inputs and 50% is used for the current layer's outputs. The small
local SRAM buffers contain the currently used weights and activations. We model
the relative energy consumption of SRAM-fetches and multiply–accumulate (MAC)
operations according to Horowitz (Horowitz 2014). Here, the energy consumption
of a read/write from/to the small local SRAM E_L is modeled to be equal to the
energy of a single MAC-operation E_{MAC}, while accessing the main SRAM costs
$E_M = 2 \times E_{\text{MAC}}$. Other operations, such as bias additions, quantized-ReLU,
and nonlinear batch normalization, are modeled to cost E_{MAC} as well. All these
numbers incorporate control, data transfer, and clocking overheads. The total on-
chip energy per inference is then the sum of the compute energy E_C and the cost
of weight E_W and activation accesses E_A. In absolute numbers $E_{\text{MAC}} = 3.7$ pJ
for a 16-bit operation, as measured in a 28 nm technology in (Moons et al. 2017c).
Energy consumption for a read/write operation to the SRAM buffers is modeled as
$E_{SL} = 2E_{\text{MAC}}$ and $E_{SM} = 10E_{\text{MAC}}$, according to (Horowitz 2014). The global
energy consumption is:

$$E_{HW} = E_C + E_W + E_A$$
$$E_C = E_{MAC}(Q) \times N_c + E_{MAC}(16) \times 3 \times A_s$$
$$E_W = E_M \times N_s + E_L \times N_c/\sqrt{p}$$
$$E_A = 2 \times E_M \times A_s + E_L \times N_c/\sqrt{p}$$

(3.3)

E_C is the sum of all energy consumed in partial product generation, biasing, batch normalization, and activation. This is computed through evaluating N_c, the network complexity in number of necessary MAC-operations. Biasing, batch normalization, and activation functions are performed in high precision (16b) on each output feature, hence the factor $3 \times A_s$. N_s is the model size in number of weights and biases and A_s is the total number of intermediate output features throughout the whole network. Weights are transferred once from the main to the local buffer and are then reused from there, leading to an equation for E_W. Here $\sqrt{p}$ is a reduction in memory energy, due to activation-level parallelism, as one weight is used simultaneously on $\sqrt{p}$ activations. A similar equation is derived for E_A, as activations are fetched/stored from/to the main buffer. The number of local activation fetches is reduced by $\sqrt{p}$, as one activation is simultaneously multiplied with $\sqrt{p}$ weights. The total level of parallelism p is a variable in these equations. Any on-chip memory can store a variable number of weights and activations, depending on the value of Q. A 2 Mb memory stores more than 2M 1b weights, but only 131k 16b weights. If either the weight size or feature map size exceeds the available on-chip size, communication with a larger off-chip DRAM memory will be necessary, as discussed in Sect. 3.2.1.

3.3   Test-Time Fixed-Point Neural Networks

Test-time FPNNs are networks that operate at low precision. They are pre-trained using floating-point numbers and quantized before inference according to the result of a post-training fixed-point analysis.

The work discussed in this subsection was published as Moons et al. (2016). This study was used as a starting point and inspiration for the circuit techniques and chip implementations discussed in Chaps. 4 and 5, respectively. The contributions of this specific work are threefold:

1. It shows computational precision can be scaled in several CNNs through test-time quantization of the layer's inputs and weights. The necessary *computational precision varies per architecture, per application, and even per layer* within a single CNN.
2. It shows CNNs are typically very sparse, definitely in networks using low precision. This is a characteristic that can be exploited to reduce their energy consumption, by skipping such redundant sparse computations.

3. The effect of reduced precision operating and sparse skipping on algorithmic accuracy and energy consumption is evaluated. The theoretically achievable energy-accuracy curves are derived for three popular CNN architectures in image classification.

3.3.1 Analysis and Experiments

The performance- and energy-related effects of test-time quantization are analyzed on three different network architectures for image classification. Because it is impossible to cover the entire spectrum of architectures and applications, three popular networks are chosen as representatives of small-, medium-, and large-scale architectures that differ significantly in their number of parameters:

- LeNet-5 on MNIST (Le Cun et al. 1990): LeNet-5 is a small network with two convolutional and two fully connected layers.
- CifarQuick on CIFAR-10 (Krizhevsky and Hinton 2009): CifarQuick is a medium-sized architecture with three convolutional and two fully connected layers. It has more filters per layer than LeNet-5 and operates on color images instead of grayscale images. It classifies images of the Cifar-10 data set and reaches an accuracy of 75.3%.
- AlexNet on ImageNet (Krizhevsky et al. 2012b): AlexNet is a large network with five convolutional and three fully connected layers. It reaches a top-5 accuracy of 80.0%.

For these experiments the open-source deep learning framework Caffe (Jia et al. 2014) was customized in order to be able to simulate quantization of the network's weights and inputs. All experiments are run on the validation sets of the discussed benchmarks. The relative accuracy is reported, i.e., the ratio of the accuracy after quantization and the accuracy of the original network.

3.3.2 Influence of Quantization on Classification Accuracy

As highlighted before, neural networks are fault-tolerant. Hence, the energy spent in high-precision computations does not lead to more accurate classification by the algorithm. To reduce the energy consumption of the CNN's computations, the main strategy of this work is to quantize all network weights and the inputs to its layers at test-time. Such quantization leads to a network that is only an approximation of the original network. The goal is to find out the influence of quantization on a network's accuracy and whether the effects differ significantly across network architectures.

A weight or layer input on the $[-1,1]$ interval (x) is quantized through the following function:

$$q = clip\left(\frac{round(2^{Q-1} \times x)}{2^{Q-1}}, -1, 1 - 2^{-(Q-1)}\right) \tag{3.4}$$

Before quantizing the weights and inputs, it is important to rescale them properly according to the distribution of their values onto this $[-1,1]$ interval. If there is a mismatch between the interval in which these values lie and the interval over which is quantized, the accuracy will drop, even at high precision. For this reason, all layer inputs and weights are rescaled in the network with a scalar value that corresponds to the maximal input or weight value (rounded to the next factor of 2) observed during a complete run over the validation data set. This ensures the limits of the quantization interval correspond to the limits of the data, and no bits are wasted. Mathematically, this scaling is the same as multiplying the results with a factor of 2, or shifting data in a fixed-point number representation. An operation that can be performed very cheaply in hardware.

3.3.2.1 Uniform Quantization and Per-Layer Rescaling

As a first experiment a single quantization setting is used for all layers in the network, and call this *uniform quantization*. Later, *per-layer quantization* is used, where each layer is quantized separately.

Figure 3.8a shows the relative accuracy of the three used networks as a function of the number of quantization bits. As can be seen in the figure, the relative accuracy stays equal to one for all three networks up to quantization with 19 bits, meaning that the quantized network reaches the exact same accuracy on its data set as the original

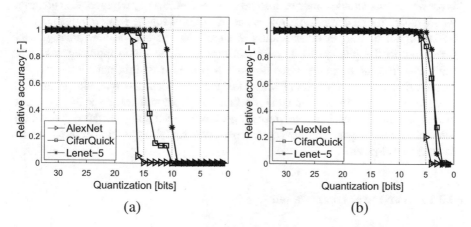

Fig. 3.8 Relative accuracy of three architectures as a function of the number of quantization bits. Layer inputs and weights are first rescaled so they can be quantized more effectively. We compare two rescaling strategies: (**a**) Uniform rescaling: all inputs and weights are scaled with the same value. (**b**) Per-layer rescaling: inputs and weights are scaled on a per-layer basis

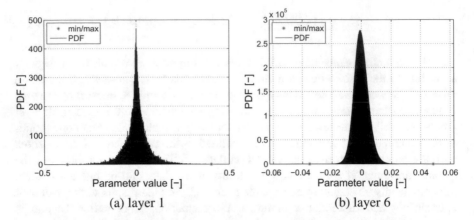

Fig. 3.9 AlexNet's first and sixth layers weight statistics. Layer 1 should be projected on the [−0.5,0.5] interval and layer 6 on the [−0.0625,0.0625] interval to minimize the needed number of bits. If layer 6 would be projected on the [−1,1] interval, four extra bits would be needed

network. At a quantization with 18-bit, however, the accuracy for AlexNet starts dropping quickly, rendering it useless for any practical application. At 11-bit the same effect can be seen for the smaller LeNet-5.

We can do better than this by applying the scaling in a more fine-grained way; we choose a different scaling factor for each layer. This is clarified by a simple example: AlexNet's first and sixth layer weight statistics are shown in Fig. 3.9. All weights of layer 1 are within the [−0.5, 0.5] interval, while the weights of layer 6 are within the [−0.0625, 0.0625] interval. By allowing to quantize the weights in layer 6 over this smaller interval instead of over [−0.5, 0.5] we recover 3 bits that would otherwise be wasted. A similar reasoning holds for the layer's inputs.

The effect of the per-layer scaling is significant, as can be seen in Fig. 3.8b. Compared to the previous, a system can now quantize much more aggressively without sacrificing accuracy. Each network performs at its original accuracy up to quantization with no more than 8 bits. After that, the accuracy again drops quickly. The reason for this improvement is the fact that input and weight statistics differ greatly among layers. Per-layer rescaling allows to set the optimal quantization interval in each layer. The concept of per-layer rescaling is also referred to as dynamic fixed-point.

3.3.2.2   Per-Layer Quantization

Just as in per-layer rescaling, a network can also have per-layer quantization: instead of quantizing all weights and inputs of the network with the same number of bits, a different setting is chosen in each layer. The idea is again to find an optimal setting for each layer by exploiting the variations of the layer-specific input and weight distributions. Another effect of the per-layer quantization is that the operating point

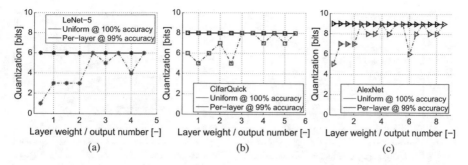

Fig. 3.10 Comparison of uniform and optimal per-layer quantization. Data points at integer values show quantization of the layer weights. Other data points show quantization of the layer inputs. Maximum achievable quantizations: (**a**) LeNet-5: 1–6 bit. (**b**) CifarQuick: 5–8 bit. (**c**) AlexNet: 5–9 bit

(i.e., the desired minimal relative accuracy) can be set more precisely. This allows controlling the energy-accuracy trade-off tightly, as discussed later in Sect. 3.3.4. This is not possible with uniform quantization: e.g., between quantization with 5 and 4 bits, the relative accuracy of LeNet-5 drops from 99.4% immediately to an unusable 86.6%.

In order to find a good quantization setting for each layer, a greedy search is performed over the parameters: starting at the first layer, its input is quantized until the accuracy drops to the target accuracy. Next, the quantization of the input is kept fixed while the quantization of its weights is maximized in the same way. The same process is applied in the next layers until the last one.

The amount of bits that can be saved with per-layer quantization at a target accuracy of 99%, compared to uniform quantization at 100% relative accuracy, is visualized for each reference network in Fig. 3.10. The results are ad-hoc, but there is a general trend of needing less bits in the lower layers of the network than in the higher layers. This is partly a result of the forward parameter sweep, but I hypothesize that the difference in input and weight statistics between lower and higher layers also plays a significant role.

3.3.3 Energy in Sparse FPNNs

This section indicates how this increased quantization can lead to energy savings in real hardware architectures. To quantify the possible energy gains of reduced precision number representations in CNN-acceleration, the energy consumption of the necessary convolutional arithmetic (multiply and add) for a full CNN algorithm is modeled according to Sect. 3.2. As this analysis was part of my earliest work, it does not yet incorporate the energy overheads of control, I/O, data and program-memory interfacing, and the clock network. It can thus be considered as the

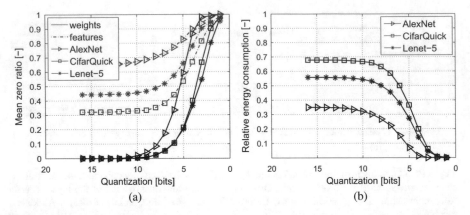

Fig. 3.11 (a) Impact of precision scaling on the average number of zeroes in the weights and input values of all three examples. (b) Relative energy reduction compared to the case without computation skipping

maximum potential energy savings through fixed-point compute. The energy model discussed above basically simplifies to $E_{DRAM} == 0$ and $n == 2$.

What is added in this analysis is the effect of sparsity in the network, as the number of zero-valued weights and inputs of typical CNNs increases at higher quantization. This can be exploited by skipping unnecessary computations with zero-valued inputs. An interesting feature of many modern convolutional neural networks is the appearance of rectified linear unit layers (ReLU layers). These put all negative inputs to zero and pass on positive values unchanged, as in Output $= max(0, Input)$. Since many layers in CNN-classification algorithms only output positive values when certain features are present, a large amount of ReLU-outputs will be zero and do not have to be used for further computations. The ReLU layers thus allow for additional energy reductions by not computing unnecessary computations through computation skipping.

Figure 3.11 shows the impact of precision scaling on the average number of zeroes in our CNN examples. For all architectures, the ratio of zeroed values lies between 50–90% of the total input values, depending on the used quantization setting. The number of zeroed values is significantly larger under precision scaling, from an average of 45% on all LeNet-5 outputs at 16-bit to 65% at 5-bit.

It is very difficult to achieve such computation skipping in a pure software solution without hardware support, since checking for zero-valued inputs in a scalar core consumes time. However, computation skipping can be achieved in a hardware accelerator with dedicated hardware support. As the algorithm can be made more sparse in turn, this is an effective way of hardware-algorithm co-optimization. In such an accelerator, flags can indicate if upcoming data is zero and prevent circuitry from switching if this is the case. As CNN-weights are fixed, these flags can be computed beforehand. An implementation of this idea is discussed in Chap. 5.

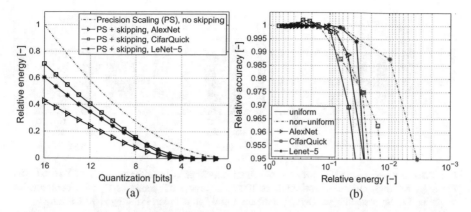

Fig. 3.12 (**a**) Relative energy consumption for uniform quantization. A striking 5–20× lower energy consumption at 8-bit, compared to the commonly used 16-bit used in the previous literature. (**b**) Energy-accuracy plot for per-layer quantization and scaling and with computation skipping. All considered CNN-networks gain at least an additional factor of 2 in energy consumption if 99% accuracy is allowed

3.3.4 Results

By combining both precision scaling and algorithmic computation skipping, major energy savings can be achieved in test-time quantized CNNs. Figure 3.12a shows the effect of uniform quantization on the energy consumption of our benchmark algorithms, if both precision scaling and computation skipping are combined. Note how the energy reduces $5 - 20\times$ for an 8-bit implementation.

Figure 3.12b shows curves in the energy-accuracy space for our CNN benchmarks. It shows the trade-off for the 95–100% classification accuracy window. All curves are given for both uniform and per-layer quantization, and are compared to the typically used uniform 16-bit number representation, without adequate precision scaling and computation skipping. If per-layer quantization is used, the trade-off is less steep than in the uniform case. This means more energy can be gained, while losing less classification accuracy in the process. All discussed networks gain at least an additional factor of 2 in energy consumption if a reduced classification accuracy of 99% is allowed.

Figure 3.13 gives an overview of the effect of each improvement on the relative energy consumption. Per-layer rescaling and computation skipping lead to the largest energy gains. Per-layer quantization leads to additional major energy gains if a reduced accuracy of 99% can be allowed.

It is important to note that the energy modeling used in this section is limited and can only be used as a first order estimation. It is not sufficient to look only at model size and the number of multiply–accumulate operations to estimate energy consumption accurately. In the next Sect. 3.4, which was part of my later work on

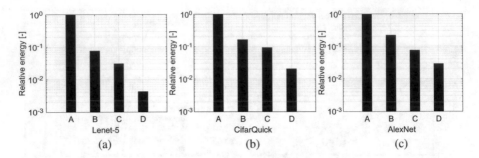

Fig. 3.13 Energy comparison between our three benchmarks in different cases. (A) 16-bit. (B) Per-layer scaling, uniform quantization at 100% accuracy. (C) Same as B, with computation skipping. Per-layer scaling, per-layer quantization and computation skipping at 99% accuracy

train-time quantized neural networks, a more accuracy energy model, based on the work in chapter refchap:envision, is used. Here, the amount of on-chip memory and the cost of DRAM is taken into account as well.

3.3.5 Discussion

By using fixed-point number approximations of floating-point baseline networks, the energy consumption in a CNN-accelerator can be reduced in two complementary ways: precision scaling and skipping of sparse computations. The combination of these two techniques leads to significant energy savings in several CNN-networks. This work shows energy reductions of up to 30× compared to the commonly used 16-bit fixed-point implementations, without sacrificing algorithm performance. If the classification accuracy can be reduced to 99%, additional energy savings can be achieved on the same network architecture, through per-layer optimization. In this case an additional 7.5× reduction can be achieved.

However, test-time quantization is sub-optimal because of two reasons. First, it offers no control over the ultimately used number of bits. This means the technique does not allow optimizing networks with a specific constrained hardware platform in mind. Second, accuracy degrades quickly at a certain cut-off point. Below this computational precision, the algorithm fails. This is illustrated in Fig. 3.14a. Here, the same network architectures are tested under different circumstances on the same benchmarks: MNIST, CIFAR-10, a custom face detection benchmark, and IMAGENET. In this figure, resulting relative performance compared to their floating-point baseline is plotted. At bitwidths lower than 4b, accuracy degrades significantly in FPNNs.

3.4 Train-Time Quantized Neural Networks

Train-time quantized neural networks (QNNs) are networks that operate at arbitrary low-precision, trained to be quantized from scratch. In train-time quantized networks, both problems of FPNNs are thereby resolved. By offering explicit control over network quantization, networks can be optimized with a specific constrained hardware platform in mind. Second, even at 1b, network topologies can be found that achieve a high accuracy, the algorithm does not necessarily fail at low precision. A full comparison between FPNN and QNN is outside the scope of this text. However, proving QNNs outperform FPNN can trivially be done by using the following experiment in Fig. 3.14. In Fig. 3.14b, the same network architectures as in Fig. 3.14a are trained from scratch as QNN. As is obvious from this Fig. 3.14, QNNs maintain higher accuracy at the same computational precision (and hence in first order energy consumption) than their FPNN counterparts at the same computational precision.

This section is primarily based on the work published in Moons et al. (2017a), which was largely built on prior work on 1b BinaryNets (Courbariaux et al. 2015; Hubara et al. 2016a,b). In this section, quantized neural networks are linked to the inference energy model from Sect. 3.2 that takes both quantization and network topology into account. Here, E_{DRAM} is taken as in Fig. 3.7b and $\epsilon ==$ 1.25, according to measurements on Envision V2 in Chap. 5. This model allows co-optimizing both the used algorithm and the hardware architecture to enable always-on embedded applications.

More specifically, my contributions in this section are the following:

- A **generalization** of the BinaryNet training setup from 1- to Q-bits (BinaryNet to intQ), where Q can be any value $\in \mathbb{N}$, both for plain CNN (Krizhevsky et al. 2012a) and ResNet (He et al. 2016) CNN-architectures.

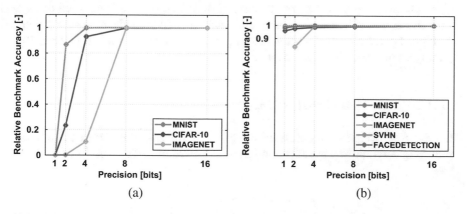

Fig. 3.14 Comparing (**a**) FPNNs versus (**b**) QNNs. Because QNNs are trained from scratch in the quantized domain, they lose less relative accuracy at lower computational precision

- An **evaluation** of the energy-accuracy-computational precision trade-off for QNN inference by linking network complexity and model size to an energy-model for a QNN chip. This setup can be used to co-optimize a neural network topology, together with the hardware platform it will run on.
- I **conclude** energy consumption at iso-accuracy varies depending on the required accuracy, computational precision, and the available on-chip memory. First, energy consumption varies orders of magnitude at iso-accuracy depending on the number of bits used in the QNN. Second, in a typical system BinaryNets or int4 operators lead to the minimum energy solution, outperforming int8 networks up to 2–10× at iso-accuracy. Third, BinaryNets are optimal in systems with tight on-chip memory constraints or at low accuracy.

3.4.1   Training QNNs

This section details my formulation of train-time quantized neural networks (QNN), which use only fixed-point representations for both weights and activations for network inference. In essence, QNNs are the generalization of binary and ternarynets (Hubara et al. 2016a; Zhu et al. 2016) to multiple bits, as in (Hubara et al. 2016b; Zhou et al. 2016).

As said before modern CPUs and GPUSs only natively support a limited number computational precision settings, such as 32–64 floating point or 8–16 bit integer operators. These machines hence don't benefit from QNNs with arbitrary lower precision. In custom designed ASICS, the used computational precision can be optimally chosen in order to minimize energy consumption per network inference at a given level of accuracy. Neurons or dot-product operators can be directly implemented as integrated digital building blocks. The complexity and energy consumption of a single neuron accumulation increases with the used number of bits Q, as illustrated in Sect. 3.2. The outputs of these neurons are always fed through a nonlinear, non-differentiable activation function which rounds these accumulated high-precision outputs to Q-bits at the output. This concept is applied in the ASIC-prototypes discussed in Chaps. 5 and 6.

3.4.1.1   Train-Time Quantized Weights

In QNNs, all weights are quantized to Q bits in a fixed-point representation. The following deterministic quantization function is used to achieve this in the forward pass:

$$q = clip\left(\frac{round(2^{Q-1} \times w)}{2^{Q-1}}, -1, 1 - 2^{-(Q-1)}\right) \tag{3.5}$$

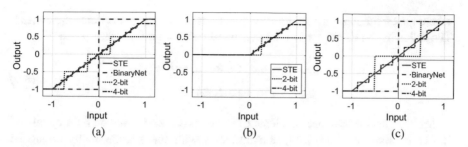

Fig. 3.15 (**a**) Weight quantization. (**b**) Quantized ReLU activation function. (**c**) Quantized hardtanh activation function. Straight-through estimators (STE) are used to estimate gradients

This is identical to the quantization in Eq. (3.4). The Q=1 case is regarded as a special case, where $q = Sign(w)$, as in the original BinaryNet paper (Hubara et al. 2016b). To successfully propagate gradients through discrete neurons, a "straight-through estimator" (STE) function is used for backpropagation. This STE leads to fastest training according to (Bengio et al. 2013). If an estimator g_q of the gradient $\frac{\partial C}{\partial q}$ has been obtained, the STE of $\frac{\partial q}{\partial w}$ is:

$$STE = hardtanh(w) = clip(w, -1, 1) \tag{3.6}$$

and the STE of $\frac{\partial C}{\partial w}$ is

$$g_w = g_q \times hardtanh(w) = g_q \times clip(w, -1, 1) \tag{3.7}$$

As in (Hubara et al. 2016a), all real-valued weights are clipped during training onto the interval $[-1, 1]$. Otherwise real-valued weights would grow large, without impacting the quantized weights. The weight quantization function $q(w)$ and STE are plotted in Fig. 3.15 for different Q.

3.4.1.2 Train-Time Quantized Activations

In QNNs, all activations are quantized to Q bits in a fixed-point representation. The following deterministic quantized activation functions are used in the forward pass:

$$A_{ReLU} = max(0, q(a))$$
$$A_{hardtanh} = qhardtanh(a) \tag{3.8}$$

where qhardtanh is as in.[1] In the BinaryNet-case, $A_{hardtanh}$ simplifies to $Sign(a)$. The following STEs are used in the backward pass for gradient estimation:

$$g_{ReLU} = max(0, hardtanh(a))$$

$$g_{hardtanh} = hardtanh(a)$$

(3.9)

Multiple setups have been evaluated. Best results are achieved with the quantized ReLU function for int2, int4, and int8 and with the symmetrically quantized hardtanh function for the Q=1 case. As in (Hubara et al. 2016a), all real valued activations are clipped during training onto the interval $[-1, 1]$. Every layer following an activation layer will then have intQ inputs. The weight quantized ReLU and hardtanh forward functions and STEs are plotted in Fig. 3.15b, c for different Q.

3.4.1.3 QNN Input Layers

In a QNN all the inputs to a layer are quantized to intQ, with the exception of the first layer, which typically has int8 pixels as its inputs. In a general case with M input bits where M>Q, an intQ layer can be performed as a sequence of shifted and added dot products. In the common case of an $M = 8$-bit input, with int4 hardware, a neuron can be computed as a series of 2 subsequent dot products, the results of which are shifted by the right amount and added together. In this case, the amount of input intQ words required to be loaded for an equivalent 8-bit representation is $M/Q = 2$.

Hence, the following encoding can be used to run M>Q inputs on intQ hardware:

$$s = x \cdot w_Q \iff s = \sum_{n=1}^{M/Q} (2^Q)^{(n-1)}(x^n \cdot w_Q)$$

(3.10)

Here, $x^{M/Q}$ is the most significant and x^1 is the least significant number of the input vector x in base Q. w_Q is a vector of weights quantized to intQ and s is the resulting weighted sum.

3.4.1.4 Quantized Training

In this work, quantized neural networks are trained from scratch, as floating-point pre-trained networks are generally an unfit initialization point for QNN training, especially for BinaryNets and int2 and int4 networks. In all experiments, training is accelerated through batch normalization (Ioffe and Szegedy 2015), which reduces

---

[1] $qhardtanh(a) = 2 \times hardtanh(round(hardsigmoid(a) \times 2^{Q-1})/2^{Q-1})$, with $hardsigmoid(a) = clip((a + 1)/2, 0, 1)$.

the overall impact of the weights' absolute scale. The ADAM learning rule (Kingma and Ba 2014) is used for weight updates, which also reduces the impact of the weight scale (Hubara et al. 2016a). All networks are trained using Theano and Lasagne for these experiments. My code in Tensorflow/Keras is available from https://github.com/BertMoons/ (Moons et al. 2017d).

3.4.2   Energy in QNNs

This section indicates how trained quantization can lead to energy savings in real hardware architectures. Contrary to the analysis of FPNNs in Sect. 3.3, this section does incorporate the energy overheads of control, I/O, data and program-memory interfacing, and the clock network. In the results given in this section, $n = 1.25$ in the model of Sect. 3.2, which is based on measurements on Envision (see Chap. 5). Also, as opposed to the analysis in Sect. 3.3, sparsity is not taken into account throughout this section. The circuitry necessary to exploit sparsity will lead to overheads at very low precision (1-4b), while it will still lead to gains when high-precision (4-16b) arithmetic is used. Adding sparsity into the energy equation for QNNs can be part of the future work in this field.

3.4.3   Experiments

3.4.3.1   Benchmarks

Several experiments on plain and ResNet-QNNs are performed for three different benchmark data sets: MNIST, SVHN, and CIFAR-10. These benchmarks are described in more detail in Sect. 1.3.4.4 of Chap. 1.

- On **MNIST** plain QNN models are trained for a maximum of 100 epochs with a continuously decaying learning rate of 0.94 starting from 0.0001. ResNet models are trained for 100 epochs, with optimal learning rate and learning rate decay depending on Q. No data augmentation is performed on this data set.
- For **SVHN** we use only 30% of the extra training set to limit training times. Plain QNN models are trained for a maximum of 150 epochs with a continuously decaying learning rate of 0.94 starting from 0.001. ResNet models are trained for 100 epochs, with optimal learning rate and learning rate decay depending on Q. No data augmentation is performed on this task.
- On **CIFAR-10**, plain QNN models are trained for a maximum of 150 epochs with a continuously decaying learning rate of 0.94 starting from 0.001. ResNet models are trained for 120 epochs, with optimal learning rate and learning rate decay depending on Q. The training set is expanded with its y-axis mirrored counterparts, but no other forms of data augmentation are applied. For testing, we only evaluate the original testing images.

Table 3.1 Used plain QNN and ResNet-QNN topologies

Block	Plain QNN	ResNet-QNN
Input	–	$16 \times 3 \times 3$
Block A – 32×32	$A \times F_A \times 3 \times 3 + \text{MaxPool}(2,2)$	$2n \times 16 \times 3 \times 3$
Block B – 16×16	$B \times F_B \times 3 \times 3 + \text{MaxPool}(2,2)$	$2n \times 32 \times 3 \times 3$
Block C – 8×8	$C \times F_C \times 3 \times 3 + \text{MaxPool}(2,2)$	$2n \times 64 \times 3 \times 3$
Output	Dense - $4 \times 4 \times F_C$ + softmax	GlobalPool + softmax

A, B, C, F_A, F_B, F_C, and n are taken as parameters. All used filters are 3×3

3.4.3.2　QNN Topologies

To quantify the energy-accuracy trade-off in QNNs, multiple network topologies are evaluated. This is necessary, as network performance not only varies with the used computational accuracy, but also with the network depth and width, hence with its computational complexity and model size. Experiments are performed on two QNN-network topologies.

- **Plain QNNs** are wide, rather than deep. Each network contains four stages: 3 QNN-blocks, each followed by a max-pooling layer and 1 fully connected classification stage as illustrated in Table 3.1. Each QNN-block is defined by two parameters: the number of basic building blocks (A, B, C) and the layer width F. Every QNN-sequence is a cascade of a QNN-layer, followed by a batch-normalization layer and a quantized activation function, as shown in Fig. 3.16a. In this work F_{Block} is varied from 32 to 512 and n_{Block} from 1 to 3. The classification stage is a sequence of a fully connected and a softmax layer. The total number of stacked weight layers is hence $n_A + n_B + n_C + 1$, with a maximum of 10.

- **ResNet-QNNs** are based on the ResNet network architecture (He et al. 2016). ResNets are deep, rather than wide, state-of-the-art networks in a number of image recognition tasks. They are considerably harder to quantize than plain QNNs, as they have a shortcut forward path that accumulates the outputs of previous layers. Each network is a stack of $6n$ layers with 3×3 convolutions on feature maps of sizes $\{s, s/2, s/4\}$ with $2n$ layers for each feature map size, with a total of $6n + 2$ stacked weighted layers. The number of filters are $\{16, 32, 64\}$ respectively as in the CIFAR-10 network in the original ResNet paper (He et al. 2016) and as illustrated in Table 3.1. The residual architecture uses basic building blocks. All QNN-layers have intQ weights and intQ inputs, but the residual path operates at full precision.

In order to reliably compare QNNs at iso-accuracy for different n, n_{Block}, F_{Block}, and Q, first the pareto-optimal floating-point architectures in the energy-accuracy space are derived. This can be done through an evolutionary architecture optimization (Real et al. 2017), but here we apply a brute search method across the parameter space. Once this pareto-optimal front is found, the same network topologies are trained again from scratch, as QNNs with a varying number of bits.

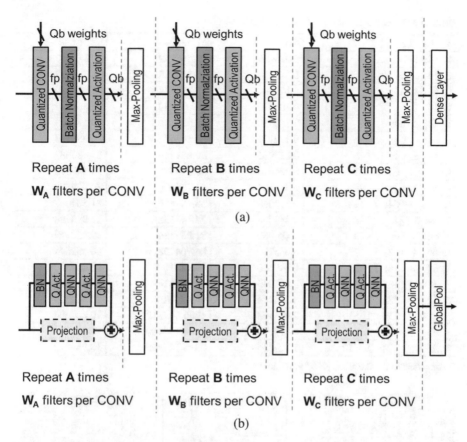

Fig. 3.16 Visual representation of benchmark networks. All networks are built from small blocks according to Table 3.1. (**a**) Plain QNN. (**b**) ResNet-QNN

3.4.4 Results

The pareto-optimal set of QNNs is analyzed in search for a minimum energy network. The list of pareto-optimal networks is given in Table 3.2 In this analysis, model parameters M_W and M_A are varied and $p = 64$. Based on measurements in (Moons et al. 2017c), we take $E_{MAC} = 3.7\,\text{pJ} \times (16/Q)^{1.25}$.

Model sizes and inference complexity are shown in Fig. 3.17. Here, computational complexity, model size, and the maximum feature map size are compared as a function of error rate and Q for the pareto-optimal plain QNN set on CIFAR-10. Figure 3.17a illustrates how the required computational complexity decreases at iso-accuracy if Q is varied from 1-to-16-bit, as networks with higher resolution require fewer and smaller neurons at the same accuracy. The int4 implementation achieves down to 9% error, while the floating-point implementation achieves down to 8.5%. At 12% error, for example, the required complexity of a float16 network is

Table 3.2 Pareto-optimal topologies on CIFAR-10

$A,\ F_A;\ B,\ F_B;\ C,\ F_C$	MWeights	MMACs	Error rate
1, 64; 1, 64; 1, 64	0.075456	13.565952	30.829327
1, 64; 1, 64; 1, 128	0.112320	15.925248	28.385417
1, 64; 1, 64; 3, 64	0.149184	18.284544	27.393830
1, 64; 1, 64; 1, 256	0.186048	20.643840	26.362179
1, 64; 2, 64; 1, 64	0.112320	23.003136	29.126603
1, 64; 2, 64; 1, 128	0.149184	25.362432	26.352163
1, 64; 2, 64; 1, 256	0.222912	30.081024	25.430689
1, 128; 2, 64; 3, 64	0.224640	38.928384	23.447516
1, 64; 3, 64; 2, 128	0.333504	44.236800	24.909856
1, 128; 3, 64; 2, 128	0.372096	55.443456	21.304087
2, 64; 1, 64; 3, 128	0.444096	72.548352	17.998798
2, 64; 2, 64; 3, 128	0.480960	81.985536	19.030449
1, 64; 2, 128; 3, 128	0.665280	86.704128	20.402644
2, 64; 3, 128; 3, 128	0.849600	162.201600	16.456330
3, 128; 2, 128; 3, 256	2.067840	475.398144	13.411458
2, 128; 3, 256; 3, 256	3.394944	645.267456	12.189503
3, 128; 3, 256; 3, 512	7.671168	1060.503552	10.840000
3, 256; 3, 256; 3, 256	4.725504	1781.268480	11.167869
3, 256; 3, 384; 3, 512	11.213568	2536.243200	10.757212

Error rates are for the BinaryNet implementations

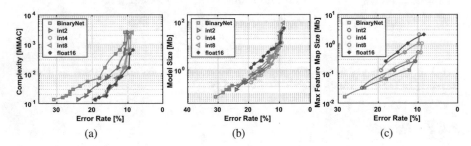

(a)   (b)   (c)

Fig. 3.17 Recognition accuracy for plain QNN networks on CIFAR-10 as a function of (**a**) computational complexity, (**b**) model size, and (**c**) maximum feature map size and the number of bits Q

80 MMAC-operations. Model complexity at iso-accuracy increases by $10\times$ to 800 MMAC-operations. On the other hand, the model size in terms of absolute storage requirements increases with the used number of bits. This is illustrated in Fig. 3.17b. Here, an int4 implementation offers the minimum model size of only 2 Mb, at 12% error rate. BinaryNets require 50% more model storage, while the float16 net requires at least $4\times$ more. Figure 3.17c shows the maximum storage required for storing intermediate feature maps as a function of network accuracy. If this size exceeds the available on-chip memory, DRAM access will be necessary. Here,

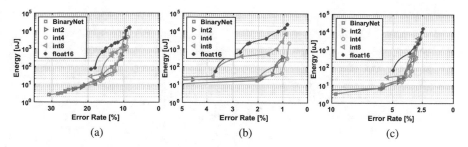

Fig. 3.18 Error rate as a function of energy consumption for a typical 4 Mb chip. (**a**) CIFAR-10. (**b**) MNIST. (**c**) SVHN

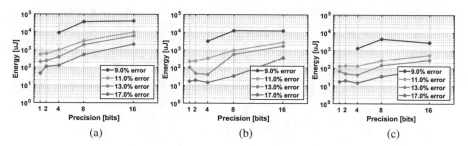

Fig. 3.19 Minimum energy plots for plain QNNs on CIFAR-10 for different amounts of on-chip memory chip-models. (**a**) 1 Mb on-chip MEM. (**b**) 4 Mb on-chip MEM. (**c**) 100 Mb on chip MEM

BinaryNets offer a clear advantage over their intQ counterparts, offering almost a 100× benefit over a float16 implementation. All plain QNN models converge at all values of Q. This is in stark contrast with (Gysel 2016), where the error rate on CIFAR-10 increases to 80% for int2 operators.

Figures 3.18 and 3.19 illustrate the **energy consumption and the minimum energy point** for plain QNN architectures. Figure 3.18 shows the error rate vs energy trade-off for different intQ implementations, for chips with a typical 4 Mb of on-chip memory. The optimal intQ mode varies with the required precision for all benchmarks. At high error rates, BinaryNets tend to be optimal. For medium and low error rates mostly int4-nets are optimal. At an error rate of 13% on CIFAR-10 in Fig. 3.18a, int4 offers a > 6× advantage over int8 and a 2× advantage over a BinaryNet. At 11%, BinaryNet is the most energy-efficient operating point and respectively 4× and 12× more energy-efficient than the int8 and float16 implementations. The same holds for networks with 10% error. However, these networks come at a 3× higher energy cost than the 11% error rate networks, which illustrates the large energy costs of increased accuracy. In an int4 network run on a 4 Mb chip, energy increase 3× going from 17% to 13%, while it increases 20× when going from 13% down to 10%. Hence, orders of magnitude of energy consumption can be saved, if the image recognition pipeline can tolerate slightly less accurate QNN architectures. Figure 3.19 compares the influence of the total on-

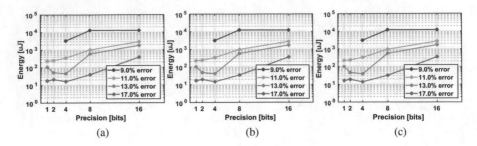

Fig. 3.20 Minimum energy plots for plain QNNs on CIFAR-10 for different amount of parallelism chip-models. (**a**) 16p. (**b**) 64p. (**c**) 256p

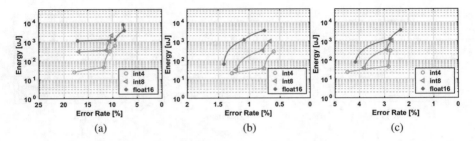

Fig. 3.21 Error rate versus energy for ResNet-QNNs. n can be $\in$ [1,3,9,18] in a $6n + 2$-layer network. (**a**) ResNet—CIFAR10. (**b**) ResNet—MNIST. (**c**) ResNet—SVHN

chip memory size $M_W + M_A$. In Fig. 3.20a, an implementation with limited on-chip memory, BinaryNets are the minimum energy solution for all accuracy-goals, as the costs of DRAM interfacing becomes dominant. In the typical case of 4 Mb, either BinaryNets, int2- or int4-networks are optimal depending on the required error rate. In a system with ∞Mb, hence without off-chip DRAM access, int2 and int4 are optimal. In all cases, int4 outperforms int8 by a factor of 2–5×, while the minimum energy point consumes 2–10× less energy than the int8 implementations.

Figure 3.21 shows the energy-vs-error trade-off for ResNet-QNNs on all benchmarks for a typical 4 Mb chip. ResNet-QNNs do not converge in the BinaryNet and int2 form, but can be trained in the int4- and int8-form. For CIFAR-10 they achieve 10% error at half the energy of their int4 plain QNN counterparts. If low error rates down to 7.5% are to be achieved on CIFAR-10, the only choice is a float16 ResNet-QNN implementation. Similarly, ResNet-QNNs achieve 1% error on MNIST at 60% of the energy consumption of the plain QNN. At a 3% error rate, int4 ResNets outperform their plain QNN counterpart by up to 6×. At the given error rates, int4 ResNet-QNNs are 5–6× more energy-efficient than their int8 counterparts at iso-accuracy for all benchmarks.

3.4.5 Discussion

This section presented a methodology to minimize the energy consumption of embedded neural networks. It is hence an example of a hardware-algorithm co-optimization method. More specifically this is done by introducing QNNs, as well as a hardware energy model used for network topology selection. To this end, the BinaryNet training setup is generalized from 1-bit to Q-bit for intQ operators, both for plain and ResNet CNN architectures. This approach allows finding the minimum energy topology and deriving several trends. First, energy consumption varies by orders of magnitudes at iso-accuracy depending on the used number of bits. The optimal minimum energy point at iso-accuracy varies between 1- and 4-bit for all tested benchmarks depending on the available on-chip memory and the required accuracy. ResNets are more energy-efficient than their iso-accuracy plain QNN counterparts, but do not converge as BinaryNets or int2-networks. BinaryNet plain QNNs are minimum energy in systems with tight on-chip memory constraints or at low benchmark accuracy. In general, int4 networks outperform int8 implementations by up to $2 - 6\times$. This suggests, the native float32/float16/int8 precision support in both low-power always-on applications and high-performance computing should be expanded with int4 to enable optimal, minimum energy inference.

3.5 Clustered Neural Networks

Linear quantization techniques offer the advantage of simplified arithmetic units in a hardware implementation. However, the energy-accuracy trade-off is not necessarily optimal for those networks, as linear quantization also reduces a network's statistical efficiency. Two nonlinear quantization techniques are analyzed in the same framework as in Sect. 3.4: clustered quantization as used in deep compression (Han et al. 2015b)(Han et al. 2016b).

In clustered quantization, a floating-point network is compressed by *clustering* its weights and activations. In this process, a set of weights or activations is k-means clustered into n clusters, such that only a few different weights are possible. Each occurrence is then described by a small index of $log_2(n)$ bits pointing to the real value in a small look-up table. After an initial clustering step, the values of these clusters are retrained to again increase the accuracy of the network. The real value is then stored as a high-precision floating-point or fixed-point number. In this case, computations are performed using complex 16b arithmetic, but the cost of all data transfers and storage are compressed. $n = 16$ is a default operating point that leads to good results.

Figure 3.22a compares the clustered approach ($\cdot$) to linear quantization ($\times$). In retrained clustered quantization, the cluster values are a mathematical optimum, rather than arbitrarily linear. This enables to maintain a higher statistical efficiency,

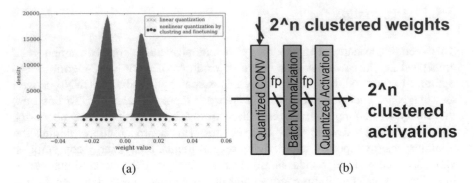

(a)   (b)

Fig. 3.22 Basic building blocks and concepts of clustered quantization. Figure is taken from (Han et al. 2016b). (**a**) Clustered vs linear quantization. (**b**) Clustered block

meaning a smaller network can achieve a higher accuracy on a benchmark. The downside of this approach is that an accurate number representation is necessary, leading to more complex arithmetic blocks than in the QNN case.

Figure 3.22b shows the basic building block of clustered neural networks. Here, both the weights and output activations of a CONV layer are clustered to n clusters. This block is then repeatedly used throughout a network, as in Fig. 3.16. As a test case, the performance of such retrained clustered network on CIFAR-10 is evaluated. This analysis clearly shows linearly quantized networks are more efficient at low accuracy (17% error rate), where models can be stored completely on chip. Clustered models are an order of magnitude more energy-efficient at high accuracy (9% error rate) than linearly quantized models (Fig. 3.23).

3.6 Conclusion

This chapter is both an introduction to various SotA means of hardware-software co-optimizations in the context of deep learning and a deeper dive in one specific method that was developed in this book: energy-efficient networks through exploiting their computational and hardware-level fault-tolerance. This is done by reducing the internal computational precision of the networks. The book distinguishes between two approaches: test-time fixed-point neural networks FPNN that are the result of an advanced fixed-point analysis on a floating-point pre-trained network and train-time quantized neural networks QNN that are trained from scratch in fixed-point. Because QNNs are explicitly trained at low precision, they outperform FPNNs in terms of energy-efficiency. It is shown optimal 4b quantized neural networks are generally optimal on the tested benchmarks, $3 - 10\times$ more efficient than 8b and 16b implementations. Moving to nonlinearly quantized neural networks, using the approach first discussed in (Han et al. 2016b), is up to $10\times$ more efficient than linear quantization at high accuracies.

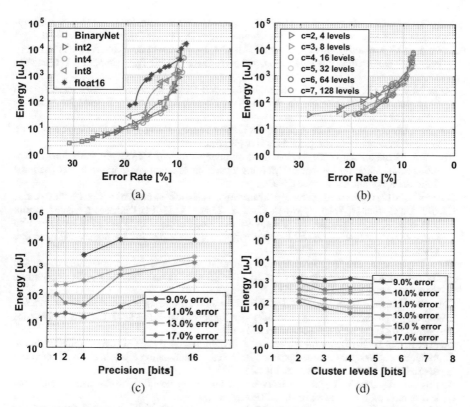

Fig. 3.23 Comparing the performance of linearly quantized versus clustered quantization on CIFAR-10. (**a**) Binary energy. (**b**) Clustered energy. (**c**) Binary minimum energy. (**d**) Clustered minimum energy

The energy gained from moving to fixed-point representations is here estimated using a high-level energy model based on the theory and measurements from Chaps. 4 and 5. Specifically for QNNs, this energy model is used to find the optimal hardware settings (the network quantization) together with the optimal network topology for a minimum energy solution.

References

Albericio J, Judd P, Jerger N, Aamodt T, Hetherington T, Moshovos A (2016) Cnvlutin:ineffectual-neuron-free deep neural network computing. In: International Symposium on Computer Architecture (ISCA)

Andri R, Cavigelli L, Rossi D, Benini L (2016) Yodann: an ultra-low power convolutional neural network accelerator based on binary weights. In: IEEE computer society annual symposium on VLSI (ISVLSI), 2016. IEEE, pp 236–241

Andri R, Cavigelli L, Rossi D, Benini L (2017) Yodann: an architecture for ultra-low power binary-weight CNN acceleration. IEEE Trans Comput Aided Des Integr Circuits Syst 37:48–60

Annaratone M, Arnould E, Gross T, Kung HT, Lam M, Menzilcioglu O, Webb JA (1987) The warp computer: architecture, implementation, and performance. IEEE Trans Comput 100(12):1523–1538

Bankman D, Yang L, Moons B, Verhelst M, Murmann B (2018) An always-on 3.8 uj/classification 86accelerator with all memory on chip in 28 nm CMOS. In: International Solid-State Circuits Conference (ISSCC) technical digest

Bengio Y, Léonard N, Courville A (2013) Estimating or propagating gradients through stochastic neurons for conditional computation. arXiv preprint:13083432

Biswas A, Chandrakasan A (2018) Conv-ram: an energy-efficient SRAM with embedded convolution computation for low-power CNN-based machine learning applications. In: International Solid-State Circuits Conference (ISSCC)

Cecconi L (2017) Optimal tiling strategy for memory bandwidth reduction for CNNS. Ph.D. thesis

Chen T, Du Z, Sun N, Wang J, Wu C, Chen Y, Temam O (2014) Diannao: a small-footprint high-throughput accelerator for ubiquitous machine-learning. In: Proceedings of the 19th international conference on architectural support for programming languages and operating systems. ACM, New York, pp 269–284

Chen YH, Emer J, Sze V (2016a) Eyeriss: a spatial architecture for energy-efficient dataflow for convolutional neural networks. In: ACM/IEEE 43rd annual International Symposium on Computer Architecture (ISCA), 2016. IEEE, pp 367–379

Chen YH, Krishna T, Emer J, Sze V (2016b) Eyeriss: an energy-efficient reconfigurable accelerator for deep convolutional neural networks. In: International Solid-State Circuits Conference (ISSCC) digest of technical papers, pp 262–263

Conti F, Benini L (2015) A ultra-low-energy convolution engine for fast brain-inspired vision in multicore clusters. In: Proceedings of the 2015 design, automation & test in Europe conference & exhibition. EDA Consortium, San Jose, pp 683–688

Courbariaux M, Bengio Y (2016) Binarynet: training deep neural networks with weights and activations constrained to +1 or -1. CoRR abs/1602.02830

Courbariaux M, Bengio Y, David JP (2015) Binaryconnect: training deep neural networks with binary weights during propagations. In: Cortes C, Lawrence ND, Lee DD, Sugiyama M, Garnett R (eds) Advances in neural information processing systems, vol 28. Curran Associates, Inc., Red Hook, pp 3123–3131

Dundar G, Rose K (1995) The effects of quantization on multilayer neural networks. IEEE Trans Neural Netw 6(6):1446–1451

Farabet C, Martini B, Corda B, Akselrod P, Culurciello E, LeCun Y (2011) Neuflow: a runtime reconfigurable dataflow processor for vision. In: IEEE computer society conference on Computer Vision and Pattern Recognition Workshops (CVPRW), 2011. IEEE, pp 109–116

Fick L, Blaauw D, Sylvester D, Skrzyniarz S, Parikh M, Fick D (2017) Analog in-memory subthreshold deep neural network accelerator. In: IEEE Custom Integrated Circuits Conference (CICC), 2017. IEEE, pp 1–4

Goetschalckx K, Moons B, Wambacq P, Verhelst M (2018) Improved deep neural network compression by combining deep compression and singular value decomposition. In: International Joint Conference on Artificial Intelligence and the 23rd European Conference on Artificial Intelligence (IJCAI-ECAI)

Gonugondla SK, Kang M, Shanbhag N (2018) A 42pj/decision 3.12tops/w robust in-memory machine learning classifier with on-chip training. In: International Solid-State Circuits Conference (ISSCC)

Gysel P (2016) Ristretto: hardware-oriented approximation of convolutional neural networks. arXiv preprint:160506402

Gysel P, Motamedi M, Ghiasi S (2016) Hardware-oriented approximation of convolutional neural networks. Workshop contribution to International Conference on Learning Representations (ICLR)

Han S, Mao H, Dally WJ (2015a) Deep compression: compressing deep neural networks with pruning, trained quantization and Huffman coding. arXiv preprint:151000149

Han S, Pool J, Tran J, Dally W (2015b) Learning both weights and connections for efficient neural network. Proceedings of advances in neural information processing systems, pp 1135–1143

Han S, Liu X, Mao H, Pu J, Pedram A, Horowitz MA, Dally WJ (2016a) EIE: Efficient Inference Engine on compressed deep neural network. In: International Symposium on Computer Architecture (ISCA)

Han S, Mao H, Dally WJ (2016b) Deep compression: compressing deep neural network with pruning, trained quantization and Huffman coding. In: International Conference on Learning Representations (ICLR)

He K, Zhang X, Ren S, Sun J (2016) Deep residual learning for image recognition. In: Conference on Computer Vision and Pattern Recognition (CVPR)

Horowitz M (2014) Energy table for 45 nm process. Stanford VLSI wiki

Huang G, Chen D, Li T, Wu F, van der Maaten L, Weinberger KQ (2017) Multi-scale dense convolutional networks for efficient prediction. arXiv preprint arXiv:170309844

Hubara I, Courbariaux M, Soudry D, El-Yaniv R, Bengio Y (2016a) Binarized neural networks. In: Advances in Neural Information Processing Systems (NIPS)

Hubara I, Courbariaux M, Soudry D, El-Yaniv R, Bengio Y (2016b) Quantized neural networks: training neural networks with low precision weights and activations. arXiv preprint:160907061

Ioffe S, Szegedy C (2015) Batch normalization: accelerating deep network training by reducing internal covariate shift. arXiv preprint:150203167

Jia Y, Shelhamer E, Donahue J, Karayev S, Long J, Girshick R, Guadarrama S, Darrell T (2014) CAFFE: Convolutional Architecture for Fast Feature Embedding. arXiv:14085093 [cs]. http://arxiv.org/abs/1408.5093, arXiv: 1408.5093

Jiang M, Gielen G (2003) The effects of quantization on multi-layer feedforward neural networks. Int J Pattern Recognit Artif Intell 17(04):637–661. https://doi.org/10.1142/S0218001403002514. http://www.worldscientific.com/doi/abs/10.1142/S0218001403002514

Jouppi NP, Young C, Patil N, Patterson D, Agrawal G, Bajwa R, Bates S, Bhatia S, Boden N, Borchers A, et al (2017) In-datacenter performance analysis of a tensor processing unit. In: International Symposium on Computer Architecture (ISCA)

Kim D, Ahn J, Yoo S (2018) Zena: zero-aware neural network accelerator. IEEE Design & Test 35:39–46

Kingma D, Ba J (2014) Adam: a method for stochastic optimization. arXiv preprint:14126980

Krizhevsky A, Hinton G (2009) Learning multiple layers of features from tiny images. Technical report

Krizhevsky A, Sutskever I, Hinton GE (2012a) Imagenet classification with deep convolutional neural networks. In: Proceedings of advances in neural information processing systems, pp 1097–1105

Krizhevsky A, Sutskever I, Hinton GE (2012b) ImageNet Classification with deep convolutional neural networks. In: Pereira F, Burges CJC, Bottou L, Weinberger KQ (eds) Advances in neural information processing systems, vol 25. Curran Associates, Inc., Red Hook, pp 1097–1105. http://papers.nips.cc/paper/4824-imagenet-classification-with-deep-convolutional-neural-networks.pdf

Le Cun BB, Denker JS, Henderson D, Howard RE, Hubbard W, Jackel LD (1990) Handwritten digit recognition with a back-propagation network. In: Advances in neural information processing systems, Citeseer

Lee J, Kim C, Kang S, Shin D, Kim S, Yoo HY (2018) Unpu: A 50.6 tops/w unified deep neural network accelerator with 1b-to-16b fully-variable weight bit-precision. In: International Solid-State Circuits Conference (ISSCC)

Lin Y, Zhang S, Shanbhag NR (2016) Variation-tolerant architectures for convolutional neural networks in the near threshold voltage regime. In: IEEE international workshop on Signal Processing Systems (SiPS), 2016. IEEE, pp 17–22

Moons B, Verhelst M (2015) DVAS: Dynamic Voltage Accuracy Scaling for increased energy-efficiency in approximate computing. In: International Symposium on Low Power Electronics and Design (ISLPED). https://doi.org/10.1109/ISLPED.2015.7273520

Moons B, Verhelst M (2016) A 0.3-2.6 tops/w precision-scalable processor for real-time large-scale convnets. In: Proceedings of the IEEE symposium on VLSI circuits, pp 178–179

Moons B, De Brabandere B, Van Gool L, Verhelst M (2016) Energy-efficient convnets through approximate computing. In: Proceedings of the IEEE Winter Conference on Applications of Computer Vision (WACV), pp 1–8

Moons B, Goetschalckx K, Van Berckelaer N, Verhelst M (2017a) Minimum energy quantized neural networks. In: Asilomar conference on signals, systems and computers

Moons B, Uytterhoeven R, Dehaene W, Verhelst M (2017b) DVAFS: Trading computational accuracy for energy through dynamic-voltage-accuracy-frequency-scaling. In: 2017 Design, Automation & Test in Europe conference & exhibition (DATE). IEEE, pp 488–493

Moons B, Uytterhoeven R, Dehaene W, Verhelst M (2017c) Envision: a 0.26-to-10 tops/w subword-parallel dynamic-voltage-accuracy-frequency-scalable convolutional neural network processor in 28 nm FDSOI. In: International Solid-State Circuits Conference (ISSCC)

Moons B, et al (2017d) Bertmoons github page. http://github.com/BertMoons. Accessed: 01 Jan 2018

Peemen M, Setio AA, Mesman B, Corporaal H (2013) Memory-centric accelerator design for convolutional neural networks. In: 2013 IEEE 31st International Conference on Computer Design (ICCD). IEEE, pp 13–19

Rastegari M, Ordonez V, Redmon J, Farhadi A (2016) XNOR-net: Imagenet classification using binary convolutional neural networks. In: European conference on computer vision. Springer, Berlin, pp 525–542

Real E, Moore S, Selle A, Saxena S, Suematsu YL, Le Q, Kurakin A (2017) Large-scale evolution of image classifiers. arXiv preprint:170301041

Shafiee A, Nag A, Muralimanohar N, Balasubramonian R, Strachan JP, Hu M, Williams RS, Srikumar V (2016) Isaac: a convolutional neural network accelerator with in-situ analog arithmetic in crossbars. In: Proceedings of the 43rd international symposium on computer architecture. IEEE Press, pp 14–26

Shin D, Lee J, Lee J, Yoo HJ (2017) 14.2 dnpu: an 8.1 tops/w reconfigurable CNN-RNN processor for general-purpose deep neural networks. In: IEEE International Solid-State Circuits Conference (ISSCC), 2017. IEEE, pp 240–241

Sze V, Yang TJ, Chen YH (2017) Designing energy-efficient convolutional neural networks using energy-aware pruning. In: Computer Vision and Pattern Recognition (CVPR)

Whatmough PN, Lee SK, Lee H, Rama S, Brooks D, Wei GY (2017) 14.3 a 28 nm soc with a 1.2 ghz 568 nj/prediction sparse deep-neural-network engine with >0.1 timing error rate tolerance for IOT applications. In: IEEE International Solid-State Circuits Conference (ISSCC), 2017. IEEE, pp 242–243

Xue J, Li J, Gong Y (2013) Restructuring of deep neural network acoustic models with singular value decomposition. In: Interspeech, pp 2365–2369

Yang L, Bankman D, Moons B, Verhelst M, Murmann B (2018) Bit error tolerance of a CIFAR-10 binarized convolutional neural network processor. In: IEEE International Symposium on Circuits and Systems (ISCAS)

Yin S, et al (2017) Minimizing area and energy of deep learning hardware design using collective low precision and structured compression. In: Asilomar conference on signals, systems and computers

Zhou S, Wu Y, Ni Z, Zhou X, Wen H, Zou Y (2016) Dorefa-net: training low bitwidth convolutional neural networks with low bitwidth gradients. arXiv preprint:160606160

Zhu C, Han S, Mao H, Dally WJ (2016) Trained ternary quantization. arXiv preprint:161201064

Chapter 4
Circuit Techniques for Approximate Computing

4.1 Introducing the Approximate Computing Paradigm

Despite significant advances in semiconductor technologies, processor architectures, and low-power design techniques, the total energy consumption of the global compute infrastructure is still rapidly increasing (Andrae 2017). Rising demands for applications in social media, augmented and virtual reality, IoT, and recognition and data-mining also lead to large increase in computational and storage demands. This trend is visible both in mobile embedded devices, used to process sensor data and perform data-mining and signal processing and recognition and in large-scale data centers that are used for storage, interconnectivity, scientific computing, social media, and financial analysis. Many of these applications are driven by the types of neural networks discussed in Chap. 1.

This increase in computational demands can come at a high economic cost. In mobile devices, there is an opportunity cost as inefficient electronics makes novel applications in AI and VR impossible. Resource constraints currently make the deployment of several IoT and recognition applications impossible, as their required computational resources do not fit in the power budget of the battery-constrained device they would run on. In data centers, energy-inefficiency leads to a real economic cost, as electricity consumption significantly increases with rising computational demands. In US data centers alone, electricity consumption alone is expected to increase from 91 billion kWh in 2013 to 140 billion kWh in 2020 (Whitney and Delforge 2014). The answer industry has offered to this challenge in the past—a mix of advances in semiconductor technologies and specialized parallel compute architectures (strategy A in the context of neural networks, see Chap. 3)—is no sufficient solution in the near future. Fortunately, many of the applications listed above are based on neural networks which show a high degree of fault-tolerance, as shown in Chap. 3. In image or speech recognition, analog signals are translated into the most likely sentences and images. As this translation is based on a series of estimates, small deviations rarely affect a user's

© Springer Nature Switzerland AG 2019
B. Moons et al., *Embedded Deep Learning*,
https://doi.org/10.1007/978-3-319-99223-5_4

satisfaction. In general a network's accuracy does not necessarily degrade under noisy computations. Other examples are in multi-media, where slight deviations in processing are likely missed by the limited perceptual capabilities of humans. Other examples are search queries, where it is more important to have a good result quickly than the best result after a long delay.

The approximate computing paradigm leverages this fault-tolerance by deliberately introducing acceptable errors into the computing process in exchange for significant gains in energy-efficiency. It can be applied on top of most existing techniques for energy minimization on multiple levels of an applications design hierarchy, from the software down to the hardware level.

An example of fault-tolerant operation was given for a series of CNNs in Chap. 3. Figure 3.10 shows the optimal number of bits in test-time FPNNs varies significantly depending on the application, the used network, and even the layer within a single CNN. Figure 3.18 shows the same for train-time QNNs. Based on this information, an optimal neural network accelerator should be able to dynamically change the number of bits it uses, depending on the requirements of the current network or application. Approximate computing techniques allow to minimize the system's energy accordingly.

Another more general example, used to illustrate the wider applicability of approximate computing methods, is given in Fig. 4.1. This figure shows the effect on image quality of low-precision JPEG compressed images. Down to 4b operators, the results of low-precision JPEG compression are barely visible to the human eye, but the operations are already 3× cheaper than an 8b ideal baseline. This shows the application's fault-tolerance and the potential to save energy by allowing (quantization) errors in the computational units. At 2b, image quality is severely degraded at an energy cost that is 5× lower than a default 8b baseline. In some applications, this can still be acceptable, if it prolongs battery life sufficiently. This example shows that in approximate computing, reconfigurability and adaptivity is of

---

The Bigger Picture

- **Chapter 1** discusses deep neural networks. They are prime examples of fault-tolerant algorithms to which approximate computing techniques can be applied.
- **Chapter 2** discusses cascaded processing hierarchies. Here, the first levels of a hierarchy typically perform simple tasks: face detection in a visual context or voice-activity detection in speech. Those are typically more tolerant to computational deviations and can be performed at low precision. More complex tasks are less tolerant to deviations in computational precision and require highly accurate computations. This "dynamic range," in necessary precision, should be supported in any modern approximate computing technique.
- **Chapter 3** shows how neural network topologies can be designed to operate using approximate computing techniques, such as precision scaling. The chapter shows the system-level effect of fixed-point operations on accuracy and energy consumption.

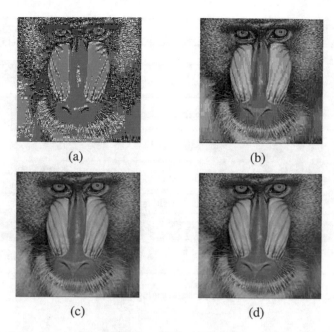

Fig. 4.1 Ideally decompressed output of a compressed JPEG baboon, where the DCT in compression is done using precision-scaled multipliers. Relative energy consumption of the MAC-units varies over $5\times$ over approximate modes at 2b to more accurate modes at 8b. Depending on a user's requirements the inaccurate modes can be used to prolong battery life. (a) 2-bit $- E_{relative} = 17\%$, (b) 4-bit $- E_{relative} = 31\%$, (c) 6-bit $- E_{relative} = 81\%$, (d) 8-bit $- E_{relative} = 120\%$

crucial importance. The output quality of any technique in approximate computing should be **dynamic** and **tunable**, in order to be able to degrade accuracy only when the application allows it.

This chapter first gives a non-exhaustive overview of different AC-techniques on all levels, based on Mittal (2016), Xu et al. (2016), Han and Orshansky (2013) in Sect. 4.2. Sections 4.3, 4.4, and 4.5 focus on my contribution to the field: dynamic-voltage-accuracy-frequency-scaling (Moons and Verhelst 2015, 2016; Moons et al. 2017b,a), a circuit-level technique with architectural implications which shows the widest energy-accuracy trade-off reported in the field.

4.2 Approximate Computing Techniques

Approximate computing techniques can be implemented on all levels of the design hierarchy of an embedded system, even though most techniques are on the circuit-level rather than on the architecture- or software-level. An overview of popular and efficient techniques in approximate computing on all levels of the design hierarchy is given in Fig. 4.2 and Sects. 4.2.4, 4.2.3, and 4.2.2.

Approximate Computing Circuits	Static Inexact arithmetic	[Liu14],[Kya11],[Kul11][Cam16], SALSA [Ven12] ASLAN[Ran14]
	Voltage over-scaling	ANT [Heg99], RAZOR [Ern03]
	Dynamic Precision Scaling	[Ven13], DVAFS [Moo17], [GS14] , [GS12], [Lee18]
Approximate Computing Software	Application Level	Loop perforation [Sid11]
	Programming Lang.	Eon [Sor07], Enerj [Sam11], Rely [Car13]
	Approx. Compilers	[Sam15]

Static Quality Prediction	Dynamic Quality Management
Error injection [Chip13], fixed point analysis [Moo16b]	SAGE [Sam13], GREEN [Bae10]

Fig. 4.2 Overview of different approximate computing methods

4.2.1 Resilience Identification and Quality Management

Even in error-tolerant applications, such as neural network acceleration, error-sensitive kernels can still exist. If inexact computations are used here, this might lead to catastrophic failure of an application. It is therefore very important to be able to formally identify and annotate parts of an application where an approximation is feasible. This resilience identification can either be determined once offline at design time, or it can be monitored continuously through dynamic quality management.

- **Resilience identification** is typically an application-dependent offline testing phase, where the impact of various computational approximations is verified for a specific application. A typical approach is to inject errors in various segments of an application and evaluate the output in a series of representative workloads (Chippa et al. 2013). Some segments, such as control-flow, are easily identified as sensitive, while others, such as iterative methods or neural networks, are easily identified as insensitive. In essence, this approach is static in the time-domain, but varies in the "spatial" algorithm domain. An example of such an analysis, in this case a fixed-point precision analysis on convolutional neural networks, is given in Moons et al. (2016) and Chap. 3. A unified, automated approach to this problem does not exist yet.
- **Quality management** is a method in which the intermediate computation quality is regularly evaluated. Based on this, it is decided if certain kernels can use approximate kernels. SAGE (Samadi et al. 2013) and GREEN (Baek and

Chilimbi 2010) compare output quality every couple of runs with a baseline golden result. Based on this they choose the approximate modes for the following computations. ANT (Hegde and Shanbhag 1999) is another example of quality management, as it uses a separate error-free unit to track and correct deviations.

4.2.2 Approximate Circuits

Several techniques for approximate computing exist on the circuit-level. These can be summarized in the following classes:

- **Inexact arithmetic circuits** are arguably the most active field of research in approximate computing. Here, arithmetic building blocks such as adders and multipliers are simplified at the circuit-level, making them inexact in some cases, but also smaller, faster, and hence more energy-efficient. This is often accomplished by simplifying a full adder design at the circuit-level, equivalent to a process altering entries in the truth table of the full adder at the functional level. Just recently, these circuits have been implemented in larger silicon prototypes (Camus et al. 2016; Moons et al. 2017b). A good overview of existing techniques for adders is given in Han and Orshansky (2013), Jiang et al. (2015). Apart from custom, hand-designed techniques, others have implemented synthesis tools that generate approximate circuitry given an accuracy constraint. SALSA (Venkataramani et al. 2012) and ASLAN (Ranjan et al. 2014) are examples that can be used to respectively synthesize combinational and sequential approximate circuitry. There are two major downsides to these techniques. First, the energy-accuracy trade-off they offer is **static** and can only be changed at design time. After an approximate arithmetic block is synthesized, its function is fixed, as is its computational accuracy and error rates. As many approximate applications and architectures require a dynamic trade-off, inexact arithmetic circuits do not offer a general solution. Second, the dynamic range of the energy-accuracy trade-off these circuits offer is limited and typically outperformed by the precision scaling techniques that are discussed below. At a 1% RMS error rate in a digital multiplier, for example, the precision scaling technique in Moons et al. (2017a) outperforms (Kyaw et al. 2011) by 5×.
- **Voltage over-scaling (VOS) techniques** let circuits operate at a frequency that is too high for the operating voltage. This removes all guard-bands and margins that are typically present in digital designs, but also induce timing errors in the computations. As digital power consumption scales quadratically with supply voltage, this technique has the potential to offer large energy gains. Moreover, the voltage or frequency in a system can be easily modulated dynamically, allowing for a dynamic energy-accuracy trade-off. As errors generated by voltage-over-scaling techniques can be catastrophically large and as they are

hard to predict, they are typically combined with architecture-level techniques such as algorithmic noise tolerance (ANT) (Hegde and Shanbhag 1999). More recent and established methods are RAZOR-setups as in Ernst et al. (2003), de la Guia Solaz and Conway (2014) that can be applied in a more general context, in microcontrollers, for example. However, these techniques can only be considered approximate computing if they end up deliberately introducing errors into the circuitry. In razor, all errors are typically corrected.

- **Precision scaling techniques** are arguably the most powerful, general, and easily available means of approximate computing. In precision scaling, the bit width of the computational operators is adapted at run-time depending on the needs of an application. Several implementations have been made for specific cases (Park et al. 2010), DSP vector processors (Venkataramani et al. 2013; de la Guia Solaz et al. 2012) with RAZOR (de la Guia Solaz and Conway 2014). This concept is taken further in this book, by combining it with voltage (Moons and Verhelst 2015) and frequency scaling (Moons et al. 2017a). These techniques are the focus of the rest of this chapter and are applied in several silicon prototypes in Chaps. 5 and 6.

4.2.3 Approximate Architectures

Designing approximate computing systems also requires adapting the used compute architectures, both to the software level on top and to the circuit-level the architecture is built on. Changes are required both on the processor-architecture level and on the SOC-architecture level.

- **Processor architectures** can be optimized for fine-grained and coarse-grained approximate computing. In fine-grained approximate computing, a specific instruction set architecture (ISA) allows a compiler to convey what can and cannot be approximated. For specific tasks, the compiler can then decide whether to map them on exact or approximate hardware. In coarse-grained approximate computing, specific code segments are offloaded to dedicated approximate accelerators or full cores outside of the processor pipeline. The potential of the first approach (Sampson et al. 2011) is very limited in a von-Neumann processor architecture, where most of the energy consumption is consumed in control, data transfers and clock distribution. None of these dominant blocks can gain from approximation, as should follow from the resilience identification process.
- Several circuit techniques for approximate computing require changes on the **SOC architecture** level. Specifically when local voltage-scaling techniques are used, such as in Venkataramani et al. (2013) and DVAFS (Moons et al. 2017a) (Sect. 4.3), the SOC has to be organized in specific voltage domains, which can heavily influence the lay-out of the full SOC and puts extra constraints on frequency generators and voltage regulators.

- In algorithmic noise tolerance (ANT) (Hegde and Shanbhag 1999, 2001) an error-free replica is put next to a main computing block as an error correcting block.
- The chip discussed in Chap. 6 is an architectural implementation of the DVAFS precision scaling concept discussed in Sect. 4.3. In this chip, rather than performing voltage and frequency scaling on individual arithmetic building blocks at lower precision, it supports sub-net parallel operation of convolutional neural networks for image recognition together with voltage and frequency scaling. It is hence an example of a system architecture tailored to the approximate computing philosophy.

4.2.4 Approximate Software

Several efforts have been made on the software level towards approximate computing.

- Approximations can be either on the **application level** through techniques such as iterative loop perforation (Sidiroglou-Douskos et al. 2011).
- Other efforts have been on **programming languages and compilers**. Programming languages such as Eon (Sorber et al. 2007), EnerJ (Sampson et al. 2011), and Rely (Carbin et al. 2013) expose approximation to the programmer through language syntax. In Sampson et al. (2011), Sorber et al. (2007) the designer should know which kernels are tolerant to approximation, while in Carbin et al. (2013) the compiler automatically derives how errors can affect algorithm performance. Such a compiler could then automatically decide which operations can be performed on approximate hardware or which operands can be approximated (Misailovic et al. 2014). The techniques are generally in support of approximate computing circuits, which is the abstraction level where most gains are expected.

4.2.5 Discussion

The different techniques discussed above in this section are all used in some form throughout this work. An example of resilience identification is given in the form of an advanced fixed-point analysis in Sect. 3.3 and to some extent below in Sect. 4.3.2. Chapter 5 focuses on the design and implementation of approximate hardware architectures using the approximate circuit techniques discussed in Sect. 4.3. These techniques are the main focus of the rest of this chapter.

4.3 DVAFS: Dynamic-Voltage-Accuracy-Frequency-Scaling as a Means for Circuit-Level Approximate Computing

An effective approach to **dynamically** scale the power consumption of, e.g., a digital multiplier at constant throughput is through at run-time truncation or rounding of a variable number of its input bits (Venkataramani et al. 2013). This introduces deviations in output quality caused by higher quantization errors on the inputs and thus decreases computational accuracy, but also reduces the circuit's internal switching activity and hence energy consumption. This effect makes it an approximate computing technique by definition.

4.3.1 DVAFS Basics

4.3.1.1 Introducing the DVAFS Energy-Accuracy Trade-Off

Figure 4.3 shows how precision scaling influences the performance of a 16b multiplier. Figure 4.3a shows the maximum- and RMS-error as a function of the used number of bits. A 5-bit rounded multiplier achieves 1.8% relative RMS error compared to an ideal solution. The introduced errors are hence small on average, while the maximum error is deterministic and limited by the used number of bits. Figure 4.3b shows the associated energy consumption of the same multiplier in different operating modes. At the same 1.8% relative RMSE point, it can consume up to $20\times$ less than the default 16b mode. This is in sharp contrast to other work (Liu et al. 2014; Kulkarni et al. 2011), where the error probability is much lower than in the DVAFS case, but the error amplitude can be large and unpredictable.

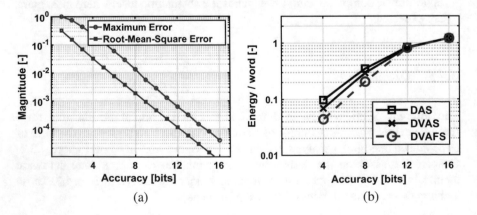

Fig. 4.3 A precision-scaled 16b multiplier in the energy-accuracy space. (**a**) RMSE as a function of number of bits, (**b**) Relative energy

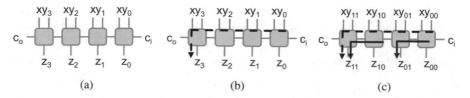

Fig. 4.4 A 4b adder example under different forms of precision scaling. **(a) DAS.** When 2 MSBs are used, the other inputs are gated. In this case only the green blocks will switch and the switching activity drops. **(b) DVAS.** When 2 MSBs are used, the switching activity drops and the critical path is shortened. **(c) DVAFS.** When 2 MSBs or less are used, redundant logic is reused to make the multiplier subword-parallel. Here, both the critical path and the switching activity are reduced. If throughput is kept constant, the operating frequency can be reduced as well

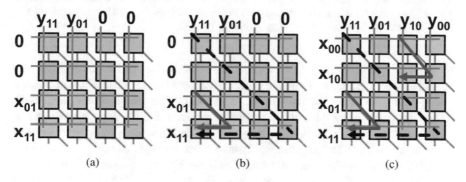

Fig. 4.5 A 4b naive carry save multiplier example. **(a) DAS.** When 2 MSBs are used, the other inputs are gated. In this case only the green blocks will switch and the switching activity drops. **(b) DVAS.** When 2 MSBs are used, the switching activity drops and the critical path is shortened. **(c) DVAFS.** When 2 MSBs or less are used, redundant logic is reused to make the multiplier subword-parallel. Here, both the critical path and the switching activity are reduced. If throughput is kept constant, the operating frequency can be reduced as well

This section further discusses dynamic-voltage-accuracy-frequency-scaling (DVAFS) (Moons et al. 2017a), a concept generalized from dynamic-voltage-accuracy-scaling (DVAS) (Moons and Verhelst 2015) and dynamic-accuracy-scaling (DAS). These techniques are the most efficient means of trading computational accuracy for system-level energy gains and a major contribution of this author.

4.3.1.2 Precision Scaling in DVAFS

The concept behind the three different operating modes from Fig. 4.3—DAS, DVAS, and DVAFS—is further illustrated for a digital adder in Fig. 4.4 and for a digital multiplier in Fig. 4.5.

- In **DAS** (Figs. 4.4a and 4.5a) only n MSBs are used when precision is scaled, while the LSBs are put to zero. No other parameters except for the input precision, hence the **switching activity**, can be modulated.

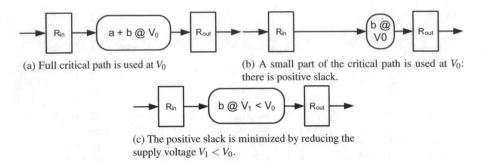

(a) Full critical path is used at V_0 (b) A small part of the critical path is used at V_0: there is positive slack.

(c) The positive slack is minimized by reducing the supply voltage $V_1 < V_0$.

Fig. 4.6 The dynamic-voltage-accuracy-scaling concept. (**a**) A combinational logic path existing out of part a and b has zero slack at supply voltage V_0. (**b**) The activity is reduced by using smaller word lengths, leading to a shorter critical path at supply voltage V_0. (**c**) The increased positive slack of the circuit allows for a lower supply voltage $V_1 < V_0$

- In **DVAS**, again only n MSBs are used. However, as the critical path has become shorter under precision scaling, the generated positive slack can be compensated for through a lower supply voltage at the same frequency. Hence, in DVAS, both the input precision related **switching activity** and the **supply voltage** can be modulated. This is further illustrated in Fig. 4.6 for a simplified general accuracy-scalable building block. Compared to the "Full critical path" mode at high precision in Fig. 4.6a, only some building blocks are switching in the low-precision mode of Fig. 4.6b. In this case, the critical path of the multiplier is also shortened from $a + b$ to b at the same voltage. This hence generates positive slack that can be compensated for through a lower supply voltage V_1, as shown in Fig. 4.6c.
- In **DVAFS**, building blocks are made subword-parallel when the required n precision drops below n_{max}. In that case, this subword-parallelism can either be converted into throughput increases, or major energy savings at constant throughput at a lower frequency. Here, the energy consumed per word is reduced considerably due to reduced **switching activity** and combined **voltage** and **frequency** scaling.

This short discussion of the three techniques is hence a first intuitive illustration of how precision scaling can be translated into energy savings on the block level. These energy gains are modeled in depth in Sect. 4.3.3 and quantified in Sect. 4.4 and Chap. 5.

4.3.2   Resilience Identification for DVAFS

Any digital system can be split into several blocks with different levels of resiliency to reduced precision operation. Not only are there different levels of resiliency, but some blocks also allow for different types of precision scaling.

Non-Accuracy-Scalable Blocks (*nas*)	Accuracy-Scalable Blocks (as)	
	Non-Voltage-Scalable Blocks (nv*as*)	Voltage-Scalable Blocks (v*as*)
All Control Blocks	Some Arithmetic, Data Transfer and Memory	Some Arithmetic
• Instruction fetch • Instruction decoding • Control logic • Instruction Memory	• Multiplexers • AND, OR, ... • Data Memory • Parallel line drivers	• Parallel Arithmetic • Multipliers • Adders

Fig. 4.7 Different categories of digital building blocks in DVAFS

Sub-blocks can be either ***accuracy-scalable*** (*as*) or ***non-accuracy-scalable*** (*nas*). A block is *as* if (1) the precision of its inputs can be modulated without breaking the functionality of the system and (2) the energy consumption of the block reduces if its input precision is scaled. If a block does not comply with requirements (1) and (2), it is considered to be *nas*. Further in this text, all *as* blocks are further subdivided into ***voltage accuracy scalable*** (v*as*) and ***non-voltage accuracy-scalable*** (nv*as*) blocks. An overview of this classification with examples is given in Fig. 4.7.

Several examples of such accuracy-scalable building blocks are given in Fig. 4.7. They are all omnipresent in today's digital designs:

1. Arithmetic blocks such as multipliers or adders are v*as*.
2. Logic blocks such as multiplexers, buffers, and drivers are nv*as*.
3. Sequential blocks such as registers and memory macros are nv*as*.

All these blocks either scale their switching activity linearly (sequential and logic blocks, adders), super-linearly, or quadratically (some multipliers) depending on their implementation.

Several other building blocks are non-accuracy-scalable *nas* in today's digital systems:

1. Control logic such as instruction decoders.
2. Control memory such as instruction SRAM.

In general, all control logic that is not operating on a system's data cannot modulate its switching activity with scaled precision.

This distinction between *as* and *nas* building blocks is crucial, as not every digital system stands to gain from the DVAFS technique. Only if a significant portion of the global power consumption is in the *as* building blocks, DVAFS will lead to large system-level gains. Hence, only signal processing systems with large, costly datapaths such as DSP processors or highly parallelized compute units with little control overhead should be targeted for the technique. General purpose CPUs or microcontrollers cannot be optimized using DAS, or approximate computing techniques in general, as typically less than 1% of their power consumption is in their arithmetic *as* parts (Horowitz 2014).

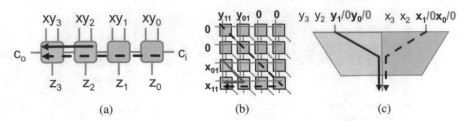

Fig. 4.8 (**a**) Adders are linear *vas*. (**b**) Multipliers are super-linear or quadratic *vas*. (**c**) Multiplexers are linear *nvas*

4.3.3 Energy Gains in DVAFS

4.3.3.1 DAS: Dynamic-Accuracy-Scaling

An example of a single multiplier operating under DAS is given in Figs. 4.8b and 4.5a. If leakage power is neglected, the power of a full DAS system, P_{DAS}, is split into an accuracy-scalable part (*as*), of which the activity is modulated with scaled precision and a non-accuracy-scalable (*nas*) part which does not modulate with scaled precision:

$$P_{DAS} = \frac{\alpha_{as}}{k_0} C_{as} f V^2 + \alpha_{nas} C_{nas} f V^2 \qquad (4.1)$$

where α is the circuit's switching activity, f is the operating frequency, C is the technology dependent switching capacitance, and V is the supply voltage. k_0 is a precision-, circuit-, and architecture-dependent scaling parameter.

4.3.3.2 DVAS: Dynamic-Voltage-Accuracy-Scaling

DVAS is a first expansion enhancing the energy-accuracy trade-off of the DAS concept. An example of a single multiplier operating under DVAS is given in Figs. 4.5b and 4.8b. As the critical paths of a DVAS building block scales under reduced precision, the generated positive slack can be compensated for through lower supply voltages, see Sect. 4.3.1.

In DVAS, these combined effects—(1) reduction of switching activity and (2) shorter critical paths allowing lower supply voltages—have a major impact on the system's dynamic power consumption. The dynamic power P_{DVAS} of a DVAS-system is split into a voltage-accuracy-scalable *vas*, an non-voltage-accuracy-scalable *nvas* part, and a non-accuracy-scalable *nas* part, given in Eq. (4.4).

$$P_{DVAS} = \frac{\alpha_{nvas}}{k_1} C_{nvas} f (V_{nvas})^2 + \frac{\alpha_{vas}}{k_2} C_{vas} f \left(\frac{V_{vas}}{k_3}\right)^2 + \alpha_{nas} C_{nas} f V_{nas}^2 \qquad (4.2)$$

This technique is analogous to widespread technologies such as clustered voltage scaling (CVS) (Usami and Horowitz 1995). In CVS, certain parts of a digital system with a slack surplus will operate at a lower supply voltage than paths without slack surplus.

4.3.3.3   DVAFS: Dynamic-Voltage-Accuracy-Frequency-Scaling

Subword-parallel DVAFS (Moons et al. 2017a,b) improves further upon the energy savings of DVAS by reusing inactive arithmetic cells at reduced precision. The example in Fig. 4.5c shows a 1–4b subword-parallel multiplier that can process two subword operations per cycle if precision is scaled to 2b or lower. In general, higher precision baselines such as 16b blocks are made reconfigurable to support parallel operation on $2 \times 8b$ or $4 \times 4b$ subwords.

If computational throughput is kept constant, this allows to drop the full system's frequency and hence its voltage significantly below DVAS values. As a result, DVAFS is the first dynamic approximate computing technique which simultaneously lowers all run-time adaptable parameters influencing power consumption: activity α, frequency f, and voltage V. In contrast to DVAS, which can only save energy in precision-scaled arithmetic blocks, DVAFS allows lowering f and V of the full system, including control units and memory, hereby shrinking even nas energy overheads drastically at low precision, see Sect. 4.4.2.

These combined effects—(1) reduction of switching activity, (2) shorter critical paths allowing lower supply voltages, (3) subword-parallel operation allowing lower operating frequencies at constant computational throughput—have a major impact on the system's power consumption. The dynamic power P_{DVAFS} at constant throughput of a DVAFS-system is given in Eq. (4.3),

$$P_{\text{DVAFS}} = \frac{\alpha_{nvas}}{k_5} C_{nvas} \frac{f}{N} (V_{nvas})^2 + \frac{\alpha_{vas}}{k_6} C_{vas} \frac{f}{N} \left(\frac{V_{vas}}{k_7} \right)^2 + \alpha_{nas} C_{nas} \frac{f}{N} \left(\frac{V_{nas}}{k_8} \right)^2$$

(4.3)

where N is the degree of subword-parallelism.

The advantage of DVAFS over DVAS is both in the extra voltage scaling capabilities it offers and in the gains in all nas blocks, such as control and memory. This can be seen from the following first order adapted formula:

$$P_{\text{DVAFS}} = \frac{N\alpha'_{nvas}}{k_1 1} C_{nvas} \frac{f}{N} (V_{nvas})^2 + \frac{N\alpha'_{vas}}{k_1 0} C_{vas} \frac{f}{N} \left(\frac{V_{vas}}{k_7} \right)^2$$
$$+ \alpha_{nas} C_{nas} \frac{f}{N} \left(\frac{V_{nas}}{k_8} \right)^2$$

(4.4)

where α_{nvas}' and α_{vas}' are the per word switching activities. As a first order analysis, the vas and as blocks see no gains from DVAFS compared to DVAS, except for the possibility of increased voltage scaling. The main gains of DVAFS are in the *nas* blocks. All control energy is scaled by N and by the factor k_5 on the voltage supply. This is important especially in systems with large *nas* building blocks, such as CPUs or DSP processors with relatively small datapaths. In DVAS, the major part of the energy consumption will have to be pushed into the *vas* and *as* building blocks, which is not always possible. If most of the energy is in *nas* blocks, only limited gains are possible through DVAS. This effect is discussed more elaborately in Sect. 4.4.2.

DVAFS is tributary to the concepts of subword-parallel processing and dynamic-voltage-frequency-scaling (DVFS) (Usami and Horowitz 1995). DVFS technique exploits the quadratic dependency of digital power dissipation to V in low-power designs. It is a system-level technique, where supply voltage is dynamically decreased or increased with the system's clock speed requirements. General purpose processors, for example, can lower their f at small workloads, when the required throughput is reduced. By doing this, slack is increased on all critical paths without inducing timing errors, allowing the supply voltage to scale accordingly. This can result in major energy savings due to the combined effect of frequency- (linear) and voltage-scaling (quadratic) on power consumption. In DVAFS however, voltage and frequency are modulated with varying accuracy rather than varying throughput requirements, making it fit within the approximate computing framework.

4.4   Performance Analysis of DVAFS

4.4.1   Block Level DVAFS

To illustrate the performance of D(V)A(F)S, the performance and energy-accuracy trade-off of a DVAFS-equipped Booth-encoded Wallace-tree multiplier is assessed. Details on the implementation of this multiplier are in Sect. 4.5.1.2. Its performance is evaluated through detailed simulations in a 40 nm *LP LVT* technology with a nominal supply voltage of 1.1 V. The multiplier is synthesized with commercially available cells in a standard digital flow, using a multi-mode optimization, ensuring the critical path indeed decreases when less bits are used. Conservative wire models are used for synthesis and power estimations. In these simulations, multiplier throughput is kept constant at $T = 1$ words/cycle$\times 500$ MHz $= 2$ words/cycle $\times 250$ MHz $= 4$ words/cycle$\times 125$ MHz $= 500$ MOPS, as shown in Fig. 4.9a.

Figure 4.9b shows the effect of D(V)A(F)S on the positive circuit delay slack. Without voltage scaling positive slack increases up to $1ns$ if 4b words are processed in the D(V)AS case and up to 7 ns due to the 125 MHz clock in the 4×4b D(V)AFS case. Figure 4.9c shows this slack can be compensated for through lowering the supply voltage at constant throughput. For DVAS, a reduction down to 0.9 V is achievable, resulting in a 36% energy reduction. In DVAFS, supply can go down to 0.75 V, or an additional decrease in energy of 55%, as shown in Fig. 4.10.

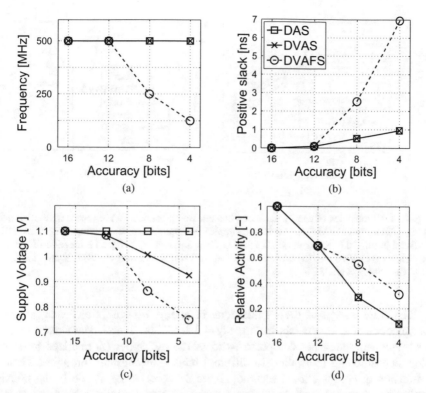

Fig. 4.9 Operating frequency, slack, supply voltage, and switching activity in a custom Booth-encoded Wallace-tree subword-parallel DVAFS multiplier in DAS-, DVAS-, and DVAFS-modes. In DVAFS, all run-time adaptable parameters in the digital power equation are modulated: α, f, and V. (**a**) Operating frequency f @ constant throughput, (**b**) Positive slack/scaling of the critical path @ 1.1 V, (**c**) Supply voltages V @ 0 positive slack, (**d**) Activity α modulation

Table 4.1 D(V)A(F)S parameters for a 16b multiplier example

Parameter	4b	8b	12b	16b
k_0	12.5	3.5	1.4	1
k_2	12.5	3.5	1.4	1
k_3	1.2	1.1	1.02	1
k_6	3.2	1.82	1.45	1
k_7	1.53	1.27	1.02	1
N	4	2	1	1

Figure 4.9a, d shows the respective operating frequencies and activity reduction of the multiplier computing at different levels of accuracy. Switching activity drops 12.5× in the DVAS-case and 3.2× in the DVAFS-case at 4b. However, in DVAFS the frequency is lowered to 125 MHz while it remains constant at 500 MHz for all accuracies in DVAS. This analysis allows extracting parameters k_i and N for this multiplier, as given in Table 4.1. Only the factors k_i for the DVAFS-compatible *vas* multiplier designed in Sect. 4.5.1.2 are given.

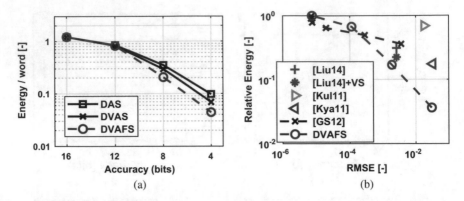

Fig. 4.10 Comparison of energy gains in a custom Booth-encoded Wallace-tree multiplier under DAS, DVAS, and DVAFS to previous examples in approximate computing (Liu et al. 2014; Kulkarni et al. 2011; Kyaw et al. 2011; de la Guia Solaz et al. 2012) as a function of the used precision. (**a**) D(V)A(F)S, (**b**) DVAFS vs (Liu et al. 2014; Kulkarni et al. 2011; Kyaw et al. 2011; de la Guia Solaz et al. 2012)

The combination of these reductions in energy consumption leads to global energy-accuracy curves shown in Fig. 4.10a. This figure compares the total energy consumption per computed word of the multiplier (normalized to a non-reconfigurable 16b multiplier) in different modes, at constant throughput, and as a function of computational accuracy. Note that processing at 16b in the DVAFS multiplier comes at a slight energy penalty, as this reconfigurability leads to a 21% overhead at full precision. At full 16b resolution the reconfigurable multiplier consumes 2.63 pJ/word compared to the 2.16 pJ/word baseline in this technology due to extra multiplexers, as discussed in Sect. 4.5.1.2. 666 The DAS case illustrates the sole effect of the decrease in α due to accuracy scaling. Energy drops more significantly in DVAS due to the added voltage scaling. DVAFS achieves energy savings of more than 95% of the baseline at 4×4 DVAFS processing.

The energy-accuracy trade-off in DVAFS is superior to other approximate computing techniques, both in terms of energy-reduction and in terms of energy-accuracy dynamic range. Figure 4.10b compares the DVAFS multiplier with the performance of Liu et al. (2014), Kulkarni et al. (2011), Kyaw et al. (2011), de la Guia Solaz et al. (2012), with energy relative to the respective fully accurate implementation in each reference and accuracy expressed in terms of root-mean-square-error (RMSE). The DVAFS-curve varies from 4b rounded operation up to 16b full precision. Although the work in de la Guia Solaz et al. (2012) consumes less energy at high accuracy, the energy consumption is higher at accuracies lower than 1e−4 RMSE. References Liu et al. (2014), Kulkarni et al. (2011), Kyaw et al. (2011) do not allow a run-time adaptable trade-off and consume more energy per word at the same level of accuracy.

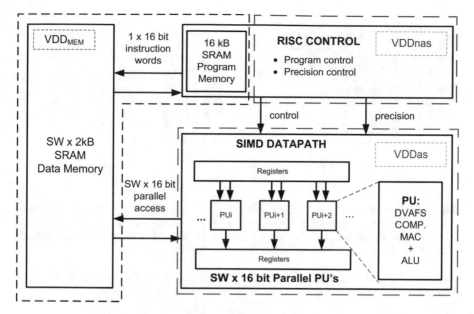

Fig. 4.11 Architecture of the parametrizable SIMD processor

4.4.2   System-Level DVAFS

To show the advantages of DVAFS at the system level, a DVAFS-compatible SIMD
RISC vector processor is implemented and simulated in an application-specific
instruction set processor (ASIP) design tool (Wu and Willems 2015), using the same
40 nm LP LVT technology as in Sect. 4.4.1. The processor, shown in Fig. 4.11,
has a parametrized SIMD width SW, denoting the number of datapath units and
memory banks. In order to be DVAFS compatible, it is subword-parallel and
contains multiple power domains. Every SIMD-unit can scale its precision across
1×1-16b, 2×1-8b, 4×1-4b DVAFS modes.

All memories are in a separate power domain (V_{MEM}), at a fixed 1.1 V to
maintain reliable operation. All other parts of the processor are in two power
domains with variable supply voltage (V_{vas} and V_{nas}). As a benchmark, a large
convolution kernel is run on the SIMD processor.

The performance of the SIMD processor is illustrated in Fig. 4.12a, where the
energy of the complete processor per processed word is plotted against the used
computational accuracy at constant throughput for processor instances with different
SIMD widths SW. The SW processor in the 1×16b mode at 500 MHZ is used as
a baseline. The maximal decrease in energy consumption is achieved at the 4×4b
DVAFS mode, with a reduction of 85% compared to the baseline. Gains are more
modest in the DAS and DVAS (60%) case as the contribution of the *as* datapath to
the full processor's energy consumption is smaller.

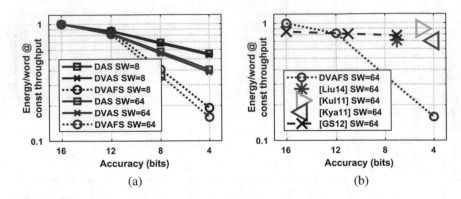

Fig. 4.12 Comparison of energy gains in a SIMD processors under DAS, DVAS, and DVAFS to previous examples in approximate computing (Liu et al. 2014; Kulkarni et al. 2011; Kyaw et al. 2011) as a function of the used precision. (**a**) DVAFS, (**b**) DVAFS vs Liu et al. (2014), Kulkarni et al. (2011), Kyaw et al. (2011), de la Guia Solaz et al. (2012)

Table 4.2 Power distribution and consumption for different DVA(F)S setups at $T = SW \times N$words/cycle$\times 500/N$ MHz. Rows in bold are DVAFS modes, plain rows are DVAS

SW	Mode	V_{nas}	V_{as}	mem	nas	as	P [mW]	P [%]
8	$1 \times 16b$	1.1 V	1.1 V	31%	46%	23%	36	100
8	$1 \times 8b$	1.1 V	1.0 V	24%	64%	12%	24	67
8	$1 \times 4b$	1.1 V	0.9 V	17%	77%	6%	20	56
8	**$2 \times 8b$**	**0.9 V**	**0.9 V**	**39%**	**48%**	**13%**	**15**	**42**
8	**$4 \times 4b$**	**0.8 V**	**0.7 V**	**47%**	**44%**	**9%**	**7**	**20**
64	$1 \times 16b$	1.1 V	1.1 V	31%	32%	37%	289	100
64	$1 \times 8b$	1.1 V	1.0 V	29%	49%	22%	160	55
64	$1 \times 4b$	1.1 V	0.9 V	23%	64%	13%	111	38
64	**$2 \times 8b$**	**0.9 V**	**0.9 V**	**41%**	**39%**	**20%**	**103**	**35**
64	**$4 \times 4b$**	**0.8 V**	**0.7 V**	**53%**	**33%**	**14%**	**45**	**16**

Voltages rounded to closest decimal point

Table 4.2 gives a more detailed overview of the energy distribution of this processor. The energy consumption is distributed over memory (*mem*), instruction decode and fetch and other overhead (*nas*), and the vector arithmetic (*vas*). The highly parallel $SW = 64$ processor has a larger percentage of its energy consumption in the arithmetic circuits and thus has a larger potential for energy reduction than its $SW = 8$ counterpart. This is the case for DAS, DVAS, and DVAFS. However, DVAFS shows a much larger energy-accuracy trade-off than the other mentioned techniques, due to its frequency modulation and improved voltage

Table 4.3 Projected energy consumption of related static approximate computing references in a $SW = 64$ SIMD processor

Ref	Eq. precision	*mem*	*nas*	*as*	P [mW]	P [%]
Liu et al. (2014)	7b	43%	45%	11%	207	71
Kulkarni et al. (2011)	5b	35%	36%	29%	257	89
Kyaw et al. (2011)	4b	44%	46%	10%	202	70
de la Guia Solaz et al. (2012)	16b	36%	38%	26%	247	85
de la Guia Solaz et al. (2012)	11b	38%	40%	23%	236	81
de la Guia Solaz et al. (2012)	7b	40%	41%	19%	226	78

The Bigger Picture

Silicon proven DVAFS implementations for energy-efficient neural network acceleration are discussed in Chaps. 5 and 6.

---

scaling. Thus, in a sole DVAS-system, the parallelism should be large for significant energy gains. In DVAFS, energy scales down significantly also at low levels of parallelism (low SW), as the supply voltage is also lowered in the non-accuracy-scalable (*nas*) parts.

Table 4.3 and Fig. 4.12b show projected gains for other recent approximate computing, as in Fig. 4.10b. This table clearly illustrates that none of the other techniques lead to significant gains on the system level, even for a highly parallel SIMD processor with $SW = 64$. On top of that, most of the other techniques are static. In de la Guia Solaz et al. (2012), the energy per processed word is brought down to 78% of the 16b baseline at an equivalent computational precision of 7b. In DVAS and DVAFS, energy is reduced down to 55% and 35% respectively in the same architecture, which is a 1.5 and 2.2 improvement over de la Guia Solaz et al. (2012).

4.5   Implementation Challenges of DVAFS

Despite the simplicity of the method, there are several challenges in implementing a well-performing DVA(F)S system, both on the functional level and in its physical implementation.

In terms of functional implementation, this section first discusses how to design basic D(V)A(F)S building blocks. Second, this section discusses the modifications necessary in order to create a working physical implementation. The reader is referred to Appendix A for details on how to design for pipe-lined circuits.

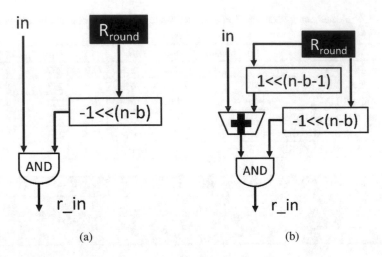

Fig. 4.13 Truncating (**a**) and rounding (**b**) a DVAS input

4.5.1   Functional Implementation of Basic DVA(F)S Building Blocks

4.5.1.1   DAS and DVAS Compatible Building Blocks

DAS and DVAS compatible building blocks can be easily implemented in any digital circuit by rounding or truncating the inputs of all *as* blocks to their N MSBs, as illustrated in Fig. 4.13. Rounding leads to zero-mean errors ($\mu = 0$, $RMSE = \sigma = \frac{LSB}{\sqrt{12}}$) but implies a significant overhead, as it requires an extra adder to compute the rounding operation. Truncating only requires programmable signal-gating circuitry but leads to higher errors with a non-zero mean ($\mu = \frac{LSB}{2}$, $RMSE = \frac{LSB}{\sqrt{3}}$).

The rounding overhead is limited only in highly parallelized architectures such as Moons and Verhelst (2016), Moons et al. (2017b). In the precision scaling used in this work, rounding units are always used. No other adaptations are needed on the RTL level to support this. The exact used precision can easily be programmed through setting the number of LSBs to be set to zero in Fig. 4.13, which can theoretically be done on a cycle by cycle basis. None of the building blocks require any changes on the RTL-level compared to a standard implementation. Hence, they can be inferred in any verilog-HDL language using the standard flows.

4.5.1.2   DVAFS-Compatible Building Blocks

In DVAFS more advanced RTL changes are required, as DVAFS makes use of subword-parallel building blocks. In any data transfer, control, or storage block, these changes are minimal or non-existent. Arithmetic blocks such as adders and

---

Tips and Tricks

Rounding (cfr. MATLAB round) requires an extra adder, but has the best numerical performance. Due to its symmetry, quantization errors are zero mean with low RMSE: $\mu = 0$, $RMSE = \sigma = \frac{LSB}{\sqrt{12}}$.

Truncating (cfr. MATLAB floor) requires no extra circuitry, but has terrible numerical performance. Due to its asymmetry, quantization errors are large with high RMSE: $\mu = \frac{LSB}{2}$, $RMSE = \frac{LSB}{\sqrt{3}}$. In any sequential algorithm, these errors will accumulate.

---

mainly multipliers do need advanced changes to make this possible. Making sure a DVAFS building block can also be used in a DVAS mode is especially challenging.

Adders can be made DVAFS compatible in a traditional way, by cutting the propagation of the carry at the desired subword-length. This can simply be implemented by gating the carry signal at the correct places, as is partly illustrated in Fig. 4.4. When processing multiple subwords at the same time all the logic released by lowering the accuracy is reused. There is hence no decrease in overall circuit activity (the switching activity per word is still lowered), but the critical path of subword adders is shorter.

Multipliers are harder to make DVAFS compatible and require a custom design. This paragraph discusses an example design of a DVAFS-compatible Booth-multiplier (Booth 1951) with a hybrid Wallace tree for partial product accumulation. Both the partial product generation and summation phases of the Booth-multiplier will have to be adapted in order to make it DVAFS compatible. A fast Brent-Kung adder (Brent and Kung 1982) is used for the final addition. A full account on how to design this multiplier can be found in Vercruysse and Uytterhoeven (2015). I acknowledge Loic Vercruysse and Roel Uytterhoeven for their hard work in building this multiplier during their master's thesis. A high-level overview of the multiplier is given in Fig. 4.14. Several changes are necessary in order to make this baseline subword-parallel while maintaining performance. More details are provided in Appendix A.

4.5.2   Physical Implementation of DVA(F)S Building Blocks

As in DVA(F)S, *nvas*, *nas*, and *vas* building blocks can operate at different voltage supplies due to the scaling of critical paths. Several architectural changes are necessary in order to implement this physically.

This poses two challenges on the physical implementation level:

- How to support **multiple supply voltages** for *nvas*, *vas*, and *nas* building blocks?
- How to **guarantee scaling of the critical path** in a real silicon physical implementation?

Fig. 4.14 All building blocks
in a DVAFS-compatible
Booth-encoded
Wallace-hybrid CSA
multiplier with a fast
Brent-Kung adder as a final
adder

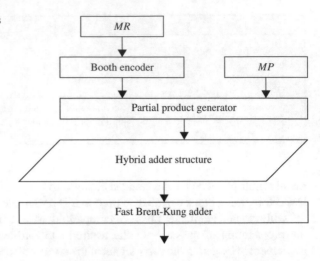

4.5.2.1   Granular Supply Scaling in DVAFS

In order to support granular DVAFS, multiple power domains are needed, all at different supply voltages. This implies adding expensive level-shifter buffers and leads to area overheads, as new power domains have to be physically split through rings in a physical implementation. Examples of how a real digital design can be split into those power domains are given in Sect. 5.3.2 of Chap. 5.

4.5.2.2   Enforcing Critical Path Scaling in DVAFS

The scaling of critical paths (with a factor k_i, as in Sect. 4.3) at reduced precision is not natively supported in digital physical implementation tools. They are not adapted to the DVAFS philosophy and only guarantee that all critical paths comply with the timing constraints. A path that has the potential to be short because of the limited number of gates in its path is not guaranteed to be short in the physical implementation: timing might be dominated by wiring. The electronic design automation (EDA) tools only guarantee it to be shorter than the timing constraints. In reality, a high percentage of paths will have a delay close to the critical path, even though logic path might be short. This phenomenon is called the *wall of slack* (Kahng et al. 2010; Pagliari et al. 2017). The DVAS idea of critical path scaling at reduced precision can still be implemented if extra (tighter) timing constraints are added for its different operating modes (Moons and Verhelst 2017).

These extra constraints have several implications. First, extra and tighter constraints lead to buffering overheads that increase the circuit's energy consumption. Second, it increases the optimization time of the back-end tool flows significantly, from 5 to 10 h for a single-constraint to 24–36 h for a multi-mode 1.5Mgate design running at 200MHz in 40nm CMOS. This could lead to a significant increase in design time that should be accounted for at the early stages in a system design.

4.6 Overview and Discussion

In this chapter, dynamic-voltage-accuracy-frequency (DVAFS) is proposed as an effective means for dynamic approximate computing on the circuit-level. This dynamic technique is compatible with deep neural networks, particularly the test-time FPNNs and train-time QNNs discussed in Chap. 3. The chapter gives an overview of the current state of the field of approximate computing and situates DVAFS in the sub-field of precision scaling. More specifically this chapter investigates the performance and implementation challenges of this technique. It is shown that, despite its simplicity, DVAFS outperforms the existing SotA in approximate computing:

1. The DVAFS concept, as well as two of its special cases: DAS and DVAS, is introduced as a means of **dynamic** approximate computing. Most other techniques in the field are **static**, meaning that their energy-accuracy trade-off cannot be modulated at run-time. This is a useful feature in neural network applications.
2. It is made clear that precision scaling techniques, among which DAS and DVAS and previous SotA techniques, only impact *accuracy-scalable as* blocks such as arithmetic and no *non-accuracy-scalable nas* blocks such as control units. This limits the techniques' applicability to very data-intensive tasks. Not so in DVAFS, a technique that also allows minimizing *nas* overheads.
3. The performance of DVAFS on the block and system level is discussed. It outperforms the static and dynamic approximate computing state of the art up to 3.5× on the block level and up to 4.5× in a parallel SIMD-processor. DVAFS offers the opportunity to lower power and energy consumption per execution for the whole system as it modulates frequency for the full system at low precision while keeping throughput constant. This allows reducing all non-compute overheads in a real approximate computing system. None of the previous works have taken this into account, making DVAFS a major contribution to the field.
4. Both the functional (subword-parallel building blocks) and physical (enforcing critical path scaling) implementation challenges of DVAFS are covered.

The processor designs covered in Chap. 5 are silicon prototypes using the DVA(F)S technique for CNNs.

References

Andrae A (2017) Consumer power consumption forecast T. In: Nordic Digital Business Summit, Helsinki
Baek W, Chilimbi TM (2010) Green: a framework for supporting energy-conscious programming using controlled approximation. In: ACM Sigplan notices, vol 45. ACM, New York, pp 198–209

Booth AD (1951) A signed binary multiplication technique. Q J Mech Appl Math 4(2):236–240

Brent RP, Kung HT (1982) A regular layout for parallel adders. IEEE Trans Comput C-31(3): 260–264

Camus V, Schlachter J, Enz C, Gautschi M, Gurkaynak FK (2016) Approximate 32-bit floating-point unit design with 53% power-area product reduction. In: 42nd European solid-state circuits conference, ESSCIRC conference 2016. IEEE, pp 465–468

Carbin M, Misailovic S, Rinard MC (2013) Verifying quantitative reliability for programs that execute on unreliable hardware. In: ACM SIGPLAN notices, vol 48. ACM, New York, pp 33–52

Chippa VK, Chakradhar ST, Roy K, Raghunathan A (2013) Analysis and characterization of inherent application resilience for approximate computing. In: Proceedings of the 50th ACM annual design automation conference, p 113

Ernst D, Kim NS, Das S, Pant S, Rao R, Pham T, Ziesler C, Blaauw D, Austin T, Flautner K, et al (2003) Razor: a low-power pipeline based on circuit-level timing speculation. In: Proceedings of the 36th annual IEEE/ACM international symposium on microarchitecture. IEEE Computer Society, Washington, DC, p 7

de la Guia Solaz M, Conway R (2014) Razor based programmable truncated multiply and accumulate, energy reduction for efficient digital signal processing. Trans VLSI syst 23: 189–193

de la Guia Solaz M, Han W, Conway R (2012) A flexible low power DSP with a programmable truncated multiplier. In: TCAS-I

Han J, Orshansky M (2013) Approximate computing: an emerging paradigm for energy-efficient design. In: 2013 18th IEEE European test symposium (ETS). IEEE, pp 1–6

Hegde R, Shanbhag NR (1999) Energy-efficient signal processing via algorithmic noise-tolerance. In: Proceedings of the 1999 international symposium on low power electronics and design. IEEE, pp 30–35

Hegde R, Shanbhag N (2001) Soft digital signal processing. IEEE Trans Very Large Scale Integr VLSI Syst 9(6):813–823

Horowitz M (2014) 1.1 computing's energy problem (and what we can do about it). In: IEEE international solid-state circuits conference (ISSCC). IEEE, pp 10–14

Jiang H, Han J, Lombardi F (2015) A comparative review and evaluation of approximate adders. In: Proceedings of the 25th edition on great lakes symposium on VLSI. ACM, New York, pp 343–348

Kahng AB, Kang S, Kumar R, Sartori J (2010) Slack redistribution for graceful degradation under voltage overscaling. In: Proceedings of the 2010 Asia and South Pacific design automation conference. IEEE Press, pp 825–831

Kulkarni P, Gupta P, Ercegovac M (2011) Trading accuracy for power with an underdesigned multiplier architecture. In: International conference on VLSI design

Kyaw KY, et al (2011) Low-power high-speed multiplier for error-tolerant application. In: Electron devices and solid-state circuits (EDSSC)

Liu C, Han J, Lombardi F (2014) A low-power, high performance approximate multiplier with configurable partial error recovery. In: Design, automation and test in Europe (DATE)

Misailovic S, Carbin M, Achour S, Qi Z, Rinard MC (2014) Chisel: reliability-and accuracy-aware optimization of approximate computational kernels. In: ACM SIGPLAN notices, vol 49. ACM, New York, pp 309–328

Mittal S (2016) A survey of techniques for approximate computing. ACM Comput Surv (CSUR) 48(4):62

Moons B, Verhelst M (2015) DVAS: dynamic voltage accuracy scaling for increased energy-efficiency in approximate computing. In: International symposium on low power electronics and design (ISLPED). https://doi.org/10.1109/ISLPED.2015.7273520

Moons B, Verhelst M (2016) A 0.3-2.6 tops/w precision-scalable processor for real-time large-scale convnets. In: Proceedings of the IEEE symposium on VLSI circuits, pp 178–179

Moons B, Verhelst M (2017) An energy-efficient precision-scalable convnet processor in 40-nm CMOS. IEEE J Solid State Circuits 52(4):903–914

Moons B, De Brabandere B, Van Gool L, Verhelst M (2016) Energy-efficient convnets through approximate computing. In: Proceedings of the IEEE winter conference on applications of computer vision (WACV), pp 1–8

Moons B, Uytterhoeven R, Dehaene W, Verhelst M (2017a) DVAFS: trading computational accuracy for energy through dynamic-voltage-accuracy-frequency-scaling. In: 2017 Design, automation & test in Europe conference & exhibition (DATE). IEEE, pp 488–493

Moons B, Uytterhoeven R, Dehaene W, Verhelst M (2017b) Envision: a 0.26-to-10 tops/w subword-parallel dynamic-voltage-accuracy-frequency-scalable convolutional neural network processor in 28nm FDSOI. In: International solid-state circuits conference (ISSCC)

Pagliari DJ, Durand Y, Coriat D, Molnos A, Beigne E, Macii E, Poncino M (2017) A methodology for the design of dynamic accuracy operators by runtime back bias. In: 2017 design, automation & test in Europe conference & exhibition (DATE). IEEE, pp 1165–1170

Park J, Choi JH, Roy K (2010) Dynamic bit-width adaptation in DCT: an approach to trade off image quality and computation energy. IEEE Trans Very Large Scale Integr VLSI Syst 18(5):787–793

Ranjan A, Raha A, Venkataramani S, Roy K, Raghunathan A (2014) ASLAN: synthesis of approximate sequential circuits. In: Proceedings of the conference on design, automation & test in Europe, European design and automation association, p 364

Samadi M, Lee J, Jamshidi DA, Hormati A, Mahlke S (2013) Sage: self-tuning approximation for graphics engines. In: 2013 46th annual IEEE/ACM international symposium on microarchitecture (MICRO). IEEE, pp 13–24

Sampson A, Dietl W, Fortuna E, Gnanapragasam D, Ceze L, Grossman D (2011) Enerj: approximate data types for safe and general low-power computation. In: ACM SIGPLAN notices, vol 46. ACM, New York, pp 164–174

Sidiroglou-Douskos S, Misailovic S, Hoffmann H, Rinard M (2011) Managing performance vs. accuracy trade-offs with loop perforation. In: Proceedings of the 19th ACM SIGSOFT symposium and the 13th European conference on foundations of software engineering. ACM, New York, pp 124–134

Sorber J, Kostadinov A, Garber M, Brennan M, Corner MD, Berger ED (2007) Eon: a language and runtime system for perpetual systems. In: Proceedings of the 5th international conference on embedded networked sensor systems. ACM, New York, pp 161–174

Usami K, Horowitz M (1995) Clustered voltage scaling technique for low-power design. In: International symposium on low power design (ISLPED)

Venkataramani S, Sabne A, Kozhikkottu V, Roy K, Raghunathan A (2012) Salsa: systematic logic synthesis of approximate circuits. In: Proceedings of the 49th annual design automation conference. ACM, New York, pp 796–801

Venkataramani S, et al (2013) Quality programmable vector processors for approximate computing. In: MICRO

Vercruysse L, Uytterhoeven R (2015) Energiewinst door good-enough computing: introductie van at run-time aanpasbare precisie in digitale circuits. PhD thesis, KU Leuven, Departement Elektrotechniek, moons, Bert and Verhelst, Marian (supervisor)

Whitney J, Delforge P (2014) Data center efficiency assessment. Issue paper on NRDC (The Natural Resource Defense Council)

Wu B, Willems M (2015) Rapid architectural exploration in designing application-specific processors. In: ASIP designer whitepaper

Xu Q, Mytkowicz T, Kim NS (2016) Approximate computing: a survey. IEEE Des Test 33(1):8–22

Chapter 5
ENVISION: Energy-Scalable Sparse Convolutional Neural Network Processing

5.1 Neural Network Acceleration

Several prior works have proposed specialized and optimized CNN dataflows for energy-efficient network inference, either on existing platforms, or on novel hardware architectures. A non-exhaustive overview is given in Sect. 3.1 in Chap. 3. Part of it is repeated here as an introduction. Optimizations for high-performance applications on CPU (Vanhoucke et al. 2011), GPU (Cavigelli et al. 2015b), or FPGA (Rahman et al. 2016; Suda et al. 2016; Motamedi et al. 2016) all consume several to hundreds of Watts, making them unusable in battery-constrained embedded systems. Other works, such as Chen et al. (2014, 2016), Knag et al. (2016), Du et al. (2015), Cavigelli et al. (2015a), Albericio et al. (2016), Han et al. (2016), Reagen et al. (2016), are ASICs which focus on low-power embedded applications, aiming to achieve real-time operation at sub-Watt power consumption. They are all accelerators, reducing their flexibility in exchange for energy-efficiency. Eyeriss (Chen et al. 2016) proposes a two-dimensional spatial architecture, exploiting data-locality and network sparsity, but does not exploit reduced precision computations. Knag et al. (2016) is an optimized architecture exploiting reduced precision, but can only perform a hardwired number of layers. DaDianNao (Chen et al. 2014) and ShiDianNao (Du et al. 2015) exploit locality, but only achieve high performance for small neural networks. Cavigelli et al. (2015a) use a specific 7×7 convolutional engine, which limits flexibility, operating at constant 12b fixed-precision. Albericio et al. (2016) combine DaDianNao's architecture with hardware support to exploit network sparsity in the time-domain, achieving up to a $1.55\times$ performance improvement. EIE (Han et al. 2016) and Minerva (Reagen et al. 2016) exploit reduced precision and sparsity in a novel hardware architecture, but tailored only for fully connected network layers.

Figure 5.1 classifies the different possible solutions for neural network acceleration in the flexibility-vs-efficiency plane. Only ASICs and application specific

© Springer Nature Switzerland AG 2019
B. Moons et al., *Embedded Deep Learning*,
https://doi.org/10.1007/978-3-319-99223-5_5

The Bigger Picture

The chips discussed in this chapter are optimized for complex CNN-based, large-scale and multi-class classification problems. Here, energy-efficiency is of utmost concern. In the hierarchical approach discussed in Chap. 2, they would hence be used in the last couple of stages of a full hierarchy. The BinarEye chips in Chap. 6 would take care of the more simple earliest stages in the hierarchy.

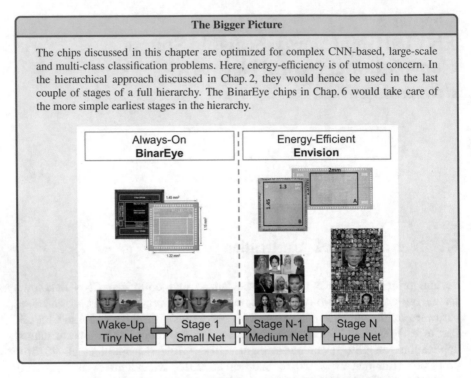

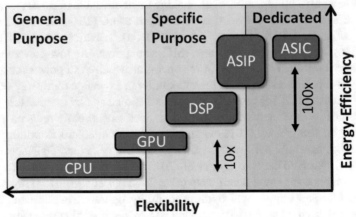

Fig. 5.1 Different hardware platforms in the energy-efficiency vs flexibility space. ASIPs offer better flexibility than hardwired or reconfigurable ASICs at a similar energy-efficiency

instruction set processors (ASIPs) (Wu and Willems 2015) are efficient enough for embedded, always-on applications.

The Envision chips (Moons and Verhelst 2016; Moons et al. 2017) discussed in this chapter are ASIP designs. They outperformed the state of the art at the time of publication by up to $5\times$ in energy-efficiency as they were the first to exploit all main energy-saving opportunities in CNNs: (a) data-locality, (b) network sparsity, and (c) precision scaling, as discussed in Chap. 1. These results were achieved through three key neural network characteristics discussed earlier in Chap. 3:

(a) A processor architecture employing a two-dimensional (2D) single instruction multiple data (SIMD) MAC array.
(b) Hardware support for network compression and guarding sparse operations. The system-level gains of this technique have been discussed for test-time FPNNs in Chap. 3.
(c) Hardware support for dynamic-voltage-accuracy-scaling (DVAS) in V1 and dynamic-voltage-accuracy-frequency-scaling (DVAFS) in V2, according to the concepts discussed in Chap. 4. We hereby illustrate the viability of the approximate computing (Chippa et al. 2013) paradigm in the context of neural networks: an efficient baseline processor can be made more energy-efficient if the algorithm requires less accurate computations.

Section 2.4 in Chap. 2 illustrates a use-case for the scalable Envision processors: a hierarchical face recognition example. This specific chapter is further organized as follows. Section 5.2 discusses the processor architecture and instruction set that is used in the Envision chips. Section 5.3 discusses the specific design and measurement results of DVAS-compatible Envision V1, implemented in a 40 nm CMOS technology. Section 5.4 discusses the design and measurements of Envision V2, implemented in a 28 nm FD-SOI technology. Finally, Sect. 5.5 explicitly compares the two versions of Envision and concludes this chapter.

5.2   A 2D-MAC Processor Architecture for Embedded CNNs

The proposed programmable and energy-efficient ASIP architecture, shown in Fig. 5.2, employs a 2D SIMD MAC array as an efficient convolutional engine (Sect. 5.2.1). For flexibility, a configurable on-chip memory architecture and on-chip direct memory access (DMA) controller (Sect. 5.2.2) have been added. Section 5.2.4 discusses the processor's instruction set.

5.2.1   Processor Datapath

5.2.1.1   2D-MAC Array

The 16×16 2D-MAC array, shown in Fig. 5.2, with its dataflow illustrated in Fig. 5.3, operates as a convolution engine. This array of single-cycle MACs achieves a $256\times$ speedup compared to a scalar solution, while minimizing bandwidth to on-chip memory. The 2D-architecture allows applying 16 different filters to 16 different

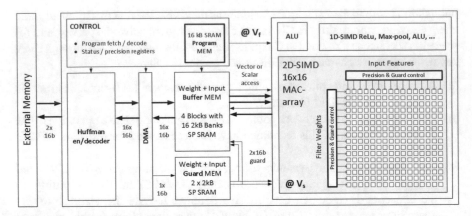

Fig. 5.2 High-level overview of the Envision architecture. The chip contains a scalar and $16\times$ vector ALU, an input and weight on-chip SP SRAM, an SRAM storing sparsity information, a control unit, a DMA, and a Huffman-based en/decoder in a fixed power domain at supply V_f. A 2D SIMD MAC array is put in a scalable power domain at supply V_s

units of the input feature map simultaneously, which is an effective exploitation of the convolutional and image reuse available in CNNs (Chen et al. 2016). This exploitation of data reuse thus allows a $256\times$ speedup, at more than $16\times$ lower internal bandwidth compared to a naive 1D-SIMD solution, requiring 2 inputs per processing unit per cycle. The local communication overhead can be further reduced by adding a FIFO register at the input of the MAC array, as shown in Fig. 5.3. Table 5.1 shows a comparison of the bandwidth fetch-reductions of the 2D-MAC array architecture compared to the naive 1D-SIMD baseline.

The example in Fig. 5.3 illustrates the four first operation steps of a typical $3 \times 3 \times C \times F$ filter dataflow, of which 4 out of F filters are performed in parallel. The concept is illustrated for a simplified 4×4 MAC array for clarity, but the chip has a 16×16 array. In this simplified example a vector is loaded from the input feature map buffer to a FIFO register as a first step. Then, it is multiplied with the first (w_{00}) filter values of four different filters out of F. All 16 partial sums are then accumulated with the previous result and stored in a matrix of local accumulation registers. In the second and third steps, a single word is fetched from the input feature map memory and pushed through the FIFO. This shifted vector is again multiplied with the next four filter values and accumulated with the previous result. This sequence is repeated three times for the 3×3 filter, illustrated by step 4, in which a vector of the next input row is multiplied and accumulated with the correct weights.

An additional advantage of this scheme is that it allows all intermediate values to be kept in the accumulation registers for a full $K \times K \times C$ convolution. Therefore the MAC accumulation registers are 48b, which is an over-design for the worst case in the AlexNet (Krizhevsky et al. 2012) benchmark. Because of this, there is no need for frequent write-backs to more expensive SRAM.

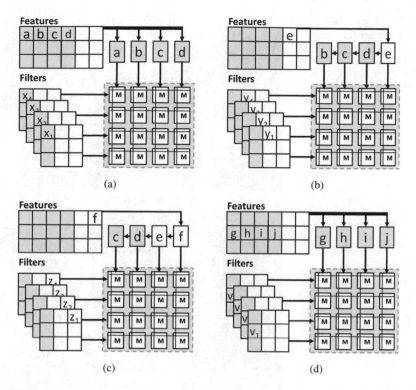

Fig. 5.3 Example of a typical CNN dataflow in Envision. (**a**) Step 1. (**b**) Step 2. (**c**) Step 3. (**d**) Step 4

Table 5.1 Words fetched per MAC-unit per MAC-operation in stride-1 operation

Filter size	1D-SIMD	2D-SIMD	2D-FIFO	Gain [×]
1×1	2	0.125	0.125	16.0
3×3	2	0.125	0.086	23.3
5×5	2	0.125	0.078	25.6
11×11	2	0.125	0.072	27.8

The presented 2D-MAC-unit with FIFO-input also supports CNN dataflows with strides not equal to 1. This is illustrated in Fig. 5.4. Figure 5.4a is a representation of the dataflow when the stride is 1 and the kernel size $k \times k$ is 3×3, similar to the flow depicted in Fig. 5.3. Figure 5.4b is a representation of a CNN-layer with $k = 3$ and a stride of 2. Here, both the input features and weights require re-ordering compared to the standard scheme with a stride of 1. In this scheme, a vector of 16 features is fetched first, together with w_{00}. This vector stores features $0, 2, \ldots, 30$ of the first line. The second cycle is a single word FIFO-access, fetching feature 32 and pushing it to the FIFO, while fetching w_{02}. Finally, this scheme requires another vector access of features $1, 3, \ldots, 31$, together with w_{01}. A final example is given

Fig. 5.4 Examples of CNN dataflow on Envision with different strides. (**a**) $k = 3$, stride = 1. (**b**) $k = 3$, stride = 2. (**c**) $k = 5$, stride = 4

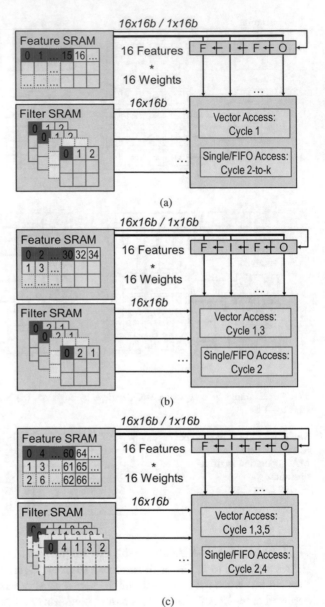

(a)

(b)

(c)

in Fig. 5.4c, where $k = 5$ and the stride is equal to 4. In this case, vector access is necessary in cycles 1, 3, and 5, while a single word FIFO-access is possible in cycles 2 and 4. In order for these schemes to be implemented, the instruction set requires sufficient flexibility in terms of vector- and word-address computations.

5.2.1.2 Other Compute Units

The processor also contains a 1D-SIMD processing unit capable of performing multiple types of vector operations. This vector unit contains 16 parallel processing units with support for bitwise- and shift operators, MAC and max-pooling (2×2 and 3×3). The processor also contains a standard scalar ALU and MAC.

5.2.2 On-Chip Memory Architecture

5.2.2.1 On-Chip Main Memory

Figure 5.5 shows the on-chip data memory topology. It is organized as one large 16b memory address space (128 kB) and is subdivided into four blocks (32 kB) containing 16 single port SRAM banks (2 kB) of 1024×16b words. Every block hence allows vector access, reading/writing one word from every bank. Single word access is also possible for scalar operations and for the FIFO-based architecture. The programmer is free to store feature maps or filter bank weights in any of the four available memory blocks.

From the processor side, two of the four blocks can be read simultaneously. Typically one block would contain only filter values and a second block a part of an input feature map. This allows fetching both filter and feature inputs for the 2D-MAC array in a single cycle. There is only one write port from the processor side, due to the lower amount of write accesses.

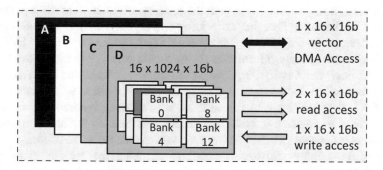

Fig. 5.5 On-chip memory architecture. One large address space is subdivided into 4 blocks, each containing 16 banks of single port SRAM. Every SRAM macro can store $1024 \times 16b$ words. The architecture allows scalar, vector, or double vector access from the processor side, while being read or written through a vector port from the DMA side

5.2.2.2 Direct Memory Access Controller

A custom designed DMA allows to efficiently communicate with off-chip memory, without stalling the processor pipeline. In parallel with access from the processor, any of the memory blocks can be read or written by the DMA which is controlled by memory-mapped registers. Synchronization between processor and DMA is done by checking DMA-specific status registers in software, indicating whether data transfer is finished.

Furthermore, it contains a specific Huffman-based en/decoder (Huffman et al. 1952) performing IO compression on sparse in- and output data (Sect. 5.2.3.2). Communication with the outside world is done using a 32b parallel interface, transferring two words per cycle if compression is turned off.

5.2.3 Hardware Support for Exploiting Network Sparsity

As discussed extensively in Sect. 3.3 in Chap. 3 in the context of test-time FPNNs, convolutional neural networks can be very sparse. This is mainly due to the use of ReLU activations and due to operation at reduced computational precision. Several hardware extensions are implemented to perform operation guarding and data compression.

5.2.3.1 Guarding Operations

At the start of a new CONV layer, both the sparsity information of the input feature map and of the layer filter banks is known. This information is stored in two dedicated, small on-chip SRAM buffers (Fig. 5.6a), containing $1024 \times 16b$ words each, storing the sparsity information for one main memory block. Every word is a $16 \times 1b$ flag, where every flag denotes the sparsity information of an associated feature map unit or filter weight. If the flag is 1, the associated word contains valid data, if it is 0 it contains zero-valued data, which can be ignored. These sparsity flags are used in Envision to both guard redundant memory fetches from memory and prevent redundant switching using zero-valued data in the MAC array.

Guarding memory fetches is done by checking the sparsity flags before performing a fetch from on-chip memory. Depending on the value of the flag, words are conditionally fetched. This is simply done by gating the enable signals to the on-chip SRAM memory banks, as in Fig. 5.6b. Only $16 + 16\ 1b$ flags are checked in order to potentially prevent 32 large SRAM banks (2 kB each) from switching. **Guarding 2D-MAC-operations** is done by both preventing the MAC inputs from switching and by clock gating the accumulation registers. To this end, extra arrays of switch-prevention circuitry, as shown in Fig. 5.6, are added. A sparsity-guard register is controlled by a sparsity flag. If the flag is 1, a new input value is stored in the register and propagated directly to the MAC array. If the flag is 0, the

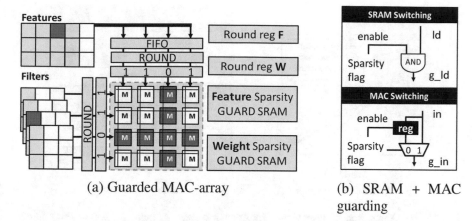

(a) Guarded MAC-array

(b) SRAM + MAC guarding

Fig. 5.6 The 2D-MAC architecture is expanded with support memory fetch- and operator-guarding to exploit network sparsity. Only the inputs of the MAC array (**a**) are guarded through arrays (**b**) guarding units.

switch-prevention register is disabled and the previous value is kept as input to the MAC array. This way, the input to a column or row of the MAC array is kept constant, effectively lowering the switching activity. The actual power reduction is dependent on the product of the sparsity levels of both the weights s_w and the feature map inputs s_i. A multiplier will only be guarded completely if both inputs are zero: $P_{rel} = s_i \times s_w$. Here, P_{rel} is the probability of a MAC to be shut down completely. s_i and s_w are the probabilities for inputs (or features) and weights to be zero.

Tips and Tricks

When guarding inputs, do not put data busses to logic 1 or 0. This might lead to additional switching activity rather than reduced switching activity, depending on the usage pattern of the guarded block. As is done here, it can be more efficient to keep data constant across cycles as a guarding technique. This way, the guarded arithmetic blocks will not switch when they are not used.

The same flags are used to clock-gate the accumulation registers. In this case, the sparsity flags are used to clock gate whole columns or rows in the MAC array at once. This leads to much higher relative gains, as only one of the inputs of the MAC must be zero to clock gate the accumulation register: $P_{rel} = s_i + s_w \times (1 - s_i)$. Here, the chance for an accumulation register to be clock-gated P_{rel} depends on the rates of input s_i and weight s_w sparsity.

The overhead of sparsity guarding is limited due to the 2D array topology. Only $32 \times 1b$ flags have to be checked in order to potentially prevent 256 MAC-units from switching.

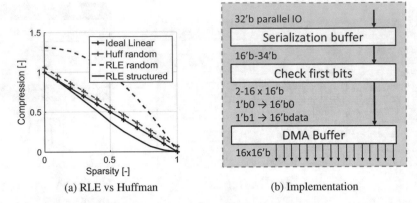

(a) RLE vs Huffman   (b) Implementation

Fig. 5.7 (**a**) Comparison of two-symbol Huffman and random, multi-symbol RLE-encoding. Huffman encoding achieves near-linear compression, independent of the data distribution. In RLE, the compression rate is a function of data-clustering and lower than two-symbol Huffman for random sparsity. (**b**) Outline of hardware implementation

5.2.3.2   Compressing IO Streams for Off-Chip Communication

Off-chip communication is controlled through a DMA containing an en/decoder. I implement a linear compression scheme based on 2-symbol Huffman encoding. If data is zero, the $16b$ data word is encoded as a $1'b0$. If the word is non-zero, it is encoded as a $17b$ word $\{1'b1, 16'bdata\}$. As shown in Fig. 5.7, this allows near ideal linear compression with the compressed data-size (C): $C = (s \times \frac{1}{n} + (1-s) \times \frac{n+1}{n})$. Here, s is the degree of sparsity and $n = 16$ is the word length.

Although other available compression algorithms such as run-length encoding (RLE), used in Chen et al. (2016) e.g., could potentially achieve super-linear compression, it does not necessarily perform well on CNN data sets. This is illustrated in Fig. 5.7a, where the plotted compression rate for RLE is a function of the data distribution and clustering, while for Huffman, it is only a function of sparsity.

An overview of the on-chip en/decoder is given in Fig. 5.7b. All words stored in on-chip memory are decompressed, as the correct data-address is crucial for pointer-based processing.

5.2.4   Energy-Efficient Flexibility Through a Custom Instruction Set

All arithmetic blocks, the memory architecture, and all precision-scaling and sparsity guarding related building blocks are controlled in the custom processor architecture shown in Fig. 5.2. This processor is built on a baseline SIMD RISC

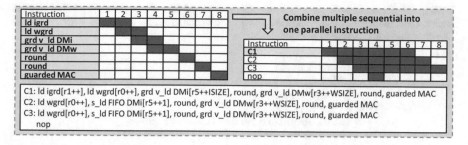

Fig. 5.8 Example assembly code for the inner loop of Listing 5.1

vector processor with a 16b instruction set and an 8192-word (16 kB) instruction memory using the Synopsys ASIP designer tool (Wu and Willems 2015). It operates a 7-stage pipeline with one fetch (IF), one decode (ID), and 5 execute (E1,...,E5) stages and is fully C-programmable using pointers. Furthermore it is equipped with numerous standard scalar- and vector-ALU instructions, as well as with jump and control instructions. Hardware loop counters are built-in for up to three nested loops.

In this work, multiple custom **variable-length instruction words** (VLIW) are added to the instruction set to, among others, support guard, round, and MAC-operations. This allows exploitation of instruction level parallelism. This is illustrated in Fig. 5.8, showing assembly code generated by the generated compiler of the ASIP design tool for the inner loop of the C-code example in Listing 5.1. This code, for a $3 \times 3 \times C$ convolution, illustrates the C-implementation of the different steps of Fig. 5.3. The listed instructions are a parallel combination of sequential sub-instructions, all of which are shorter than the standard 16b instruction length. Instead of calling these sequentially, the multiple sequential sub-instructions are combined into one single parallel instruction. The assembly instructions in Fig. 5.8, are hence the combination of 2 *guard fetch*, 2 *guarded vload*, 2 *round*, and a *guarded MAC* instruction with a short, variable length combined into one 16b instruction word. Most sub-instructions can be called individually as well. By creating these parallel instructions, the number of cycles for the inner loop of Listing 5.1 can be reduced from $3 \times 7 = 21$ to $3 + 1 = 4$.

5.2.5 Conclusion and Overview

5.3 A DVAS Compatible Envision Processor in 40 nm CMOS

The baseline processor architecture discussed in Sect. 5.2 can benefit from dynamic precision scaling by making it compatible with dynamic-voltage-accuracy-scaling (DVAS), as discussed in Chap. 4. The architecture is expanded on the RTL-level by adding functional support for precision scaling and on the physical level by

Listing 5.1 Example code for a 3×3×C CONV layer. *pI* is a vector pointer, *Ig* is the guarded vector, *Igr* is the guarded and rounded vector. The compiled assembly associated with this code is shown in Fig. 5.8. C1-3 are as in Fig. 5.8

```
for (int f = 0;f<F;f++){
  for (int h1=0;h1<H;h1++){
    for (int h2=0;h2<H;h++) {
      pW = (vint*) &Weights[f(f,h)];
      for (int c=0;c<C;c++){
        pIg = (int*) &Feature_grds[fun(f,h1,h2,c)];
        pWg = (int*) &Weight_grds[fun(f,h1,h2,c)];
        for (int k = 0; k<K;k++){
          // Fig 7. step 1, 4, 7
     |    pI = (vint*) &Features[fun(f,h1,h2,c,k)];
C1   |    Igr = round(guard(pI++,I,pIg));   // guard,round
     |    Wgr = round(guard(pW++,W,pWg));   // guard,round
     |    R = gMAC(Igr,Wgr,R,pIg++,pWg++);  // guarded mac
          // Fig 7. step 2, 5, 8
     |     pIs = (int*) &Features[fun(f,h1,h2,c,k)+1];
     |     Is = fifo(pIs++);                //FIFO ld
C2   |     Igr = round(guard(Is,I,pIg++));   // guard,round
     |     Wgr = round(guard(pW++,W,pWg++)); // guard,round
     |     R = gMAC(Igr,Wgr,R,pIg++,pWg++);  // guarded mac
          // Fig 7. step 3, 6, 9
     |     pIs = (int*) &Features[fun(f,h1,h2,c,k)+2];
     |     Is = fifo(pIs++);                //FIFO ld
C3   |     Igr = round(guard(Is,I,pIg++));   // guard,round
     |     Wgr = round(guard(pW++,W,pWg++)); // guard,round
     |     R = gMAC(Igr,Wgr,R,pIg++,pWg++);  // guarded mac
        }
      }
    }
  }
}
```

splitting the design into a scalable power domain for the MAC array and a fixed power domain for all other blocks. In order to enforce critical path scaling at lower precision, an advanced back-end place and route optimization is needed as well. This section discusses the design and measurements of DVAS compatible Envision V1, implemented in 40 nm CMOS.

5.3.1   RTL Level Hardware Support

As discussed in Sect. 4.5 in Chap. 4, DVAS compatibility requires RTL changes. The main points listed there are repeated here for completeness. The inputs to the MAC array in the scalable voltage domain are precision scaled by rounding their values to a programmable number of most-significant-bits (MSB). The least-significant-bits

Fig. 5.9 A single rounding building block exists out of one adder, two programmable shifters, and 16 AND-gates

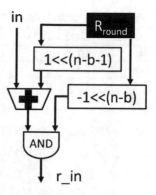

(LSB) are explicitly gated to zero, to reduce switching activity. This is done through classical round-up rounding, transforming the positive number 10010 to 10100 if it is quantized to 3b MSB. This process requires two additions to the processor architecture.

First, two programmable status registers are added, as in Fig. 5.6a, one for feature map inputs and one for filter weights, containing the number of bits that will be used for computation. These status registers can be written from software with a latency of 2 cycles. In practice, the precision is only changed on a per-layer basis. Second, two 1 × 16 vector arrays of programmable rounding units have to be added at the input of the MAC array, as shown in Fig. 5.6a. Figure 5.9 shows the building blocks of this rounding unit: one adder, two shifters, and 16 AND-gates to put the LSBs of a word to zero.

The 2D-architecture is crucial to limit the overhead of the rounding units. Only 2 × 16 = 32 inputs have to be rounded in order to provide rounded inputs to 16 × 16 = 256 MAC-units.

5.3.2 Physical Implementation

Creating a multi-power domain DVAS design, with one fixed and one scalable power domain for the MAC array, implies several complications in the physical implementation. Not only should the layout of the floorplan be adapted to contain level shifters, multiple power grids, and multiple supply ports, the back-end optimization flow should be changed as well. This is already highlighted in Chap. 4 and repeated here in the context of Envision for completeness.

DVAS is not automatically possible if the back-end optimizations are only performed in full precision mode. In a real chip, there are no guarantees that critical paths actually decrease at low computational precision, even if the circuit architecture does permit it. Theoretically shorter paths in a multiplier can still be critical at low precision due to sub-optimal placement, high wire delays or small transistor sizing. Therefore, in DVAS it is necessary to perform an advanced

Table 5.2 Multi-mode optimization settings

Mode	Fixed V [V]	Scalable f [MHz]	Scalable V [V]	Fixed f [MHz]
4b	1.1	400	0.9	200
8b	1.1	333	0.9	200
12b	1.1	200	1	200
16b	1.1	200	1.1	200

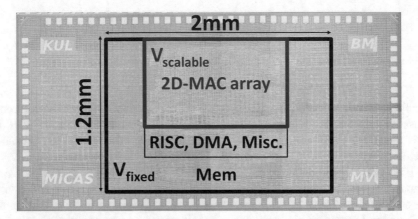

Fig. 5.10 Chip photograph and layout overview of the processor. The system totals 2.4 mm^2 in a 40 nm LP technology

multi-mode, multi-corner place, and route optimization scheme enforcing the theoretical potential of shorter critical paths at lower precision in the physical implementation. Ideally, layout should be optimized for the continuous precision range from 1 to 16 bit, using libraries characterized at all supply voltages. However, satisfactory results were achieved optimizing only for the 4-, 8-, 12-, and 16b modes. The design was optimized for 200 MHz operation in different simultaneously performed optimization modes, listed in Table 5.2. If libraries were not available at the necessary voltage, the circuit was optimized at the lowest available voltage with heavier timing constraints.

Figure 5.10 is a die photograph of the fabricated chip. The chip requires 2.4 mm^2 of active area in a 40 nm LP CMOS technology. Also indicated in this figure is the placement of the different building blocks. All DVAS-compatible building blocks, such as the 2D-MAC array data path are placed in a single power domain with a scalable supply voltage $V_{scalable}$. Another power domain contains all non-accuracy-scalable (*nas*) logic that is incompatible with DVAS voltage scaling, such as all instruction control logic, scalar ALUs, and the system's DMA. This power domain is at a fixed voltage $V_{fixed} = V_{nominal} = 1.1$ V. A third power domain contains all program and data memories and is also at the same supply voltage V_{fixed}. The domains are separated for academic purposes only, in order to be able to separately measure the power consumed by the memories.

5.3.3 Measurement Results

The performance of the Envision V1 processor in 40 nm LP was measured in the lab. Section 5.3.3.1 discusses its baseline performance. Sections 5.3.3.2 and 5.3.3.3 discuss the individual gains of DVAS and sparsity guarding. Section 5.3.3.4 demonstrates the influence of the combined techniques on a number of benchmarks.

5.3.3.1 Performance of the Full Precision Baseline

A non-unrolled $5 \times 5 \times C$ CONV layer, similar to Listing 5.1, is executed on the processor. At it's nominal supply voltage of 1.1 V and operating at full precision, the processor operates at a nominal frequency of 204 MHz and a peak performance of 204 MHz$\times$256MACs$\times 2$ $= 102$ giga-operations per second (GOPS). Every MAC-unit hence performs two operations per cycle: 1 multiply and 1 add. The real coding efficiency, the average number of MAC instructions per cycle, of the processor is typically lower and dependent on the operated filter sizes, as listed in Table 5.3. The reference code does not achieve peak performance, but has a coding-efficiency of 72%. Hence the effective performance in this case is 74 GOPS. In this nominal mode, the processor consumes 287 mW, achieving 372 peak or 270 effective GOPS/W in energy-efficiency. Table 5.4 shows a measured power breakdown of the processor. If operated simultaneously, the DMA and en/decoder consume an additional 2.5 mW, the extra on-chip SRAM consumption in this case ranges anywhere from 5 to 10 mW depending on the data-sparsity and precision.

If lower throughput is allowed, the full precision processor can easily be operated at lower voltages, as in Fig. 5.11. It runs at 100 MHz at 0.8 V, consuming 72 mW at 37 GOPS or 510 GOPS/W. Energy-efficiency improves further up until 900 GOPS/W in full precision mode, if operating at 12 MHz at 0.6 V, consuming only 5 mW at 4.4 GOPS.

Table 5.3 MAC-efficiency as a function of filter sizes

Filter size	Not-unrolled [%]	Unrolled [%]
1×1	33	50
3×3	53	75
5×5	72	83
11×11	85	92

Table 5.4 Full precision power breakdown at 1.1 V and 204 MHz

	Dynamic [mW]	Leakage [mW]	Total [mW]
Program Mem	4.1	0.4	22.5
Data Mem	18		
Control	6.4	0.3	18.7
Data Transfer	12		
MAC array	244	1.6	245.6
Total	284.5	2.3	286.8

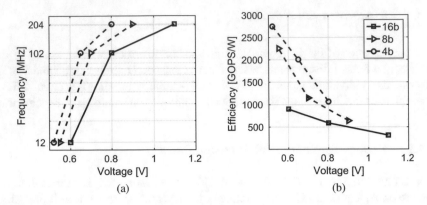

Fig. 5.11 (**a**) Frequency and (**b**) efficiency versus supply voltage in different modes. In the 4*b* and 8*b* modes, the fixed power domain is running at a higher supply voltage. This voltage is the same as the one in the 16*b* mode at a given frequency

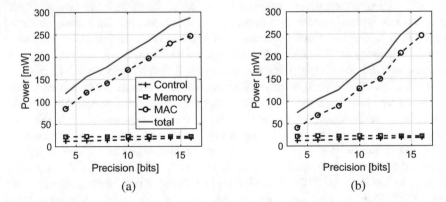

Fig. 5.12 Effect of precision scaling on power consumption under (**a**) DAS and (**b**) DVAS

5.3.3.2   Performance Under Dynamic Precision DVAS

In DVAS the supply of the scalable power domain can be lowered at constant frequency, as illustrated in Fig. 5.11 for different modes. The fixed power domain is always operating at the supply necessary for full precision operation. At 204 MHz, the scalable supply can be lowered down to 0.9 V and 0.8 V for 8b and 4b operation, respectively. Up to 2715 GOPS/W operation is measured at 12 MHz and 4b precision. Figure 5.12 shows the explicit effect of precision scaling and voltage scaling on the power consumption of different parts of the processor. The measured efficiency gains are significant: in a typical AlexNet l2 (Table 5.5), reducing precision down to 7b improves energy-efficiency 2.6× compared to the full precision baseline.

Table 5.5 Performance comparison of relevant benchmarks, running at 204 MHz

Layer	Weight bits	Input bits	Weight sparsity (%)	Input sparsity (%)	Weight BW Reduc.	Input BW reduc.	IO (MB/f)	HuffIO (MB/f)	Voltage (V)	MMACs/ frame	Power (mW)	Effective TOPS/W
General CNN	16	16	0%	0%	1.0×	1.0×	–	–	1.1	–	**287**	**0.3**
AlexNet l1	7	4	21%	29%	1.17×	1.3×	1	0.77	0.85	105	85	0.96
AlexNet l2	7	7	19%	89%	1.15×	5.8×	3.2	1.1	0.9	224	55	1.4
AlexNet l3	8	9	11%	82%	1.05×	4.1×	6.5	2.8	0.92	150	77	0.7
AlexNet l4	9	8	04%	72%	1.00×	2.9×	5.4	3.2	0.92	112	95	0.56
AlexNet l5	9	8	04%	72%	1.00×	2.9×	3.7	2.1	0.92	75	95	0.56
Total/avg.	–	–	–	–	–	–	19.8	**10**	–	–	**76**	**0.9**
LeNet-5 l1	3	1	35%	87%	1.40×	5.2×	0.003	0.001	0.7	0.3	25	1.07
LeNet-5 l2	4	6	26%	55%	1.25×	1.9×	0.050	0.042	0.8	1.6	35	1.75
Total/avg.	–	–	–	–	–	–	0.053	**0.043**	–	–	**33**	**1.6**

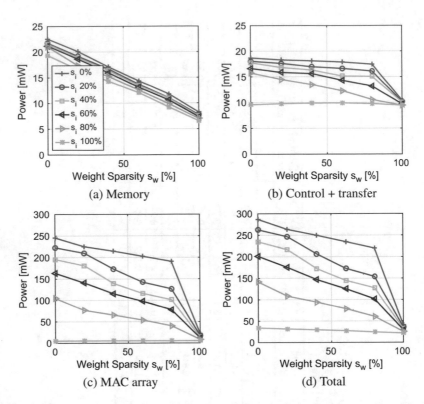

Fig. 5.13 Influence of input and weight sparsity guarding on the power consumption of the (**a**) Memory, (**b**) control unit and data transfer, (**c**) the 2D-MAC array and (**d**) gives an overview of the global power consumption

5.3.3.3   Performance on Sparse Datastreams

Figure 5.13 shows the individual effects of sparsity guarding in different parts of the processor, using the same reference code as in the previous sections at nominal frequency and supply voltage. Memory energy consumption scales down linearly with the sparsity degrees s_w and s_i. The effect of s_i on memory power consumption is limited, because the FIFO-based architecture requires less reads from the input feature memory than from the weight memory. The global energy gains are small if only one of the inputs is sparse. Still, in a typical AlexNet l3 (Table 5.5), relatively high sparsity $s_w = 11\%$ and $s_i = 82\%$ leads to a large $3\times$ energy decrease in the MAC array and a $2.4\times$ decrease for the whole system at full precision.

5.3.3.4   Performance on Benchmarks

Table 5.5 shows the combined effect of the proposed techniques on the energy consumption of three benchmarks: AlexNet (Krizhevsky et al. 2012), Lenet-5

Fig. 5.14 Influence of the different techniques on the power consumption of AlexNet layer 2. A total 5× gain in power consumption is achieved through precision, voltage scaling, and sparse operator guarding

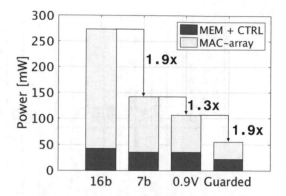

(LeCun et al. 1998), and the full precision baseline reference layer of Sect. 5.3.3.1. All measurements are again performed at the nominal 204 MHz.

As shown in Table 5.5 and Fig. 5.14, the filter weights and feature map inputs can be represented using only 7b. Reducing global power consumption by 1.9× from 274 mW at full precision down to 142 mW. Voltage in the MAC array can be set to 0.9 V, lowering power by an additional 1.3× to 107 mW. With these quantization settings, sparsity in the filters is 20% and 89% in the input feature map. If operations and memory fetches are guarded as well, power consumption goes down another 1.9× to 55 mW, which is a total 5× gain compared to the already efficient 2D-baseline processor architecture. Also, because of sparsity, the IO-communication can be compressed up to 5.8× for input features. Similar results are illustrated for the other layers in Table 5.5, resulting in an average of 76 mW or 0.9 TOPS/W for AlexNet, running at 47 fps.

LeNet-5 is more sparse and requires less computational precision, even down to 1b in the first layer, gaining another factor of 2× in energy-efficiency. The combination of these effects—lower precision arithmetic, lower voltages, and more guarded operations—allows running LeNet-5 at 13k fps at an efficiency of 1.6 effective TOPS/W or 2.5 μJ/frame. We hereby illustrate the flexibility, performance, and unique energy-scalability of this design.

5.3.3.5 Comparison with the State of the Art

As shown in Table 5.6, CPU and GPU implementations (Cavigelli et al. 2015b) are extremely flexible but consume >100W at low energy-efficiency. Origami (Cavigelli et al. 2015a), running at 12b, achieves an energy-efficiency of 437 GOPS/W, but cannot scale its energy consumption depending on the application's requirements. Knag et al. (2016) is a hardwired ASIC for a fixed two-layer network topology. It only achieves high performance when the network is >90% sparse. Eyeriss (Chen et al. 2016), implemented in a 65 nm CMOS technology and running at 16b, consumes 278 mW or 166 GOPS/W on the AlexNet benchmark at a throughput of 34.7 fps on the CONV layers, or an efficiency of 8 mJ/frame.

Table 5.6 Comparison of this work with previous published CNN implementations

Reference	DAC'15 (Cavigelli et al. 2015b)	DAC'15 (Cavigelli et al. 2015b)	GLSVLSI'15 (Cavigelli et al. 2015a)	VLSI'16 (Knag et al. 2016)	ISSCC'16 (Chen et al. 2016)	This work (Moons and Verhelst 2016)
Technology	CPU[a]	GPU[b]	65 nm	40 nm	65 nm LP	40 nm LP
Gate count [NND2]	-	-	912k	-	1852k	1600k
Core area [mm^2]	-	-	1.31	1.41	12.25	2.4
On-chip SRAM [kB]	-	-	43	52 (Reg)	181.5	144
# MAC-units [-]	-	-	196	2432 (Eq.)	168	256
Frequency [MHz]	3700	852	500	240	200	204
Peak perf. [GOPS]	118	365	196	898	67	102
Avg. perf. [GOPS]	35	84	145	-	60	74
Bit width [bits]	32 float	32 float	12 fixed	8 fixed	16 fixed	1–16 progr.
Filters [-]	All	All	< 7 × 7	< 8 × 8	All	All
Channels [-]	All	All	1–256	3, 16	1–1024	All
Filters [-]	All	All	1–256	16, 64	1–1024	All
Layers [-]	All	All	All	2	All	All
Stride [-]	All	All	-	No	All	1–4 [h], all [v]
Power @ f_{nom} [mW]	130,000	11,000	510	141	235–332	25–300
Energy [GOPS/W]	0.15	8.6	437–803	6400	160–250	270–2750
(AlexNet) [mW]	-	-	-	-	278	76
(AlexNet) [fps]	-	-	-	-	34.7	47

[a] E5-1620v2
[b]Tegra K1

This work achieves 47 fps throughput on the AlexNet CONV layers at nominal speed, consuming 76 mW or 1.6 mJ/frame if the system is fully optimized. This is hence a 5× improvement in energy-delay-product (EDP) over the AlexNet-benchmarked reference. The processor further allows scaling energy-efficiency depending on the network's requirements. LeNet-5 consumes only 25–33 mW or 1600 GOPS/W at the same nominal 204 MHz clock frequency, due to DVAS and sparsity guarding. No other work allows such network dependent energy-scalability at nominal throughput.

5.3.4 Envision V1 Overview

A DVAFS precision-scalable processor for CNNs with operator- and memory-guarding was fabricated in a 40 nm LP CMOS technology. It has a total active area of 2.4 mm^2 and runs at a nominal frequency of 204 MHz at 1.1 V.

First, the processor is fully C-programmable and uses a 256 2D-MAC array architecture as an efficient convolution baseline. This 2D-architecture allows inherent 16× reuse of filter weights and feature map inputs. A FIFO further reduces internal memory bandwidth up to 27.8×.

Second, the processor minimizes energy by modulating precision and supply voltage in the MAC array from layer-to-layer. This requires extra rounding circuitry, a split into two power domains and an advanced back-end place and route optimization. When scaling down from 16- to 8- or 4b at 204 MHz, the scalable supply voltage can go down from 1.1 V nominally to 0.9 V and 0.8 V, respectively. In AlexNet layer 2, this leads to a 2.6× gain compared to the full precision baseline. Only 32 rounding units are necessary to supply rounded inputs to 256 MACs.

Finally, sparsity in CNNs is exploited through preventing SRAM banks and the 2D array from switching and by compressing data up to 5.8×. In AlexNet layer 2, sparsity leads to a 2× gain compared to the precision and voltage scaled baseline, while only 32 sparsity flags are required to potentially prevent 32 SRAM banks and 256 MACs from switching.

Because of these techniques this chip minimizes energy consumption for any CNN, showing up to 5× improvement over the prior state of the art in terms of energy/frame. We hereby enable low-power, high-performance applications of computer vision for battery powered devices.

However, Envision V1 was only DVAS compatible. As shown in Chap. 4, the energy-scalability of the same architecture can be further improved upon by making it dynamic-voltage-accuracy-frequency-scaling (DVAFS) compatible. This technique leads to larger energy savings especially in low-precision operating points, as it also reduces non-compute overheads that might become dominant. Section 5.4 discusses the design and measurements of DVAFS-compatible Envision V2.

5.4   A DVAFS-Compatible Envision Processor in 28 nm FD-SOI

The baseline processor architecture discussed in Sect. 5.2 can benefit further from dynamic precision scaling by making it DVAFS compatible. As discussed in Chap. 4, a DVAFS-compatible system uses subword-parallel building blocks at low precision. A 16b multiplier, for example, can be used to process $2\times$ 8b words if such low computational precision is allowed by the algorithm. If in this case system-level throughput is kept constant, all run-time adaptable variables α, f, V in the digital power equation $P = \alpha C f V^2$ can be modulated to save in energy consumption. The major advantage over DVAS, which only allows α and V modulation at constant throughput, is that DVAFS allows minimizing non-compute overheads as well.

A conceptual overview of the different supported operating modes and their influence on the processing architecture is given in Fig. 5.15. The same 2D-MAC architecture as in Envision V1 is used. However, Envision V2 supports 3 operating modes:

1. A default DVAS compatible 1-16b operating mode, as in Envision V1. Every memory address occupies 1 1-16b word and every MAC processes 1 word per cycle.
2. A $2\times$ subword-parallel mode can be used when precision drops below 8b. Every memory address then contains $2\times$ 1-8b words while every MAC processes 2 words per cycle, an effective $2\times$ throughput increase at a fixed frequency.
3. A $4\times$ subword-parallel mode can be used when precision drops below 4b. Every memory address then contains $4\times$ 1-4b words while every MAC processes 4 words per cycle, an effective $4\times$ throughput increase at a fixed frequency.

In modes (2) and (3), the chip effectively uses all three ways of reuse available in CNNs: convolutional, image, and filter reuse (Chen et al. 2016). In Envision V1, only the convolutional and image reuse can be exploited. This section further discusses the design and measurements of DVAFS-compatible Envision V2, implemented in 28 nm FD-SOI.

5.4.1   RTL Level Hardware Support

As DVAFS-compatible building blocks have to be made subword-parallel, several RTL changes have to be made to the baseline architecture. A detailed discussion on the adaptations necessary to build a subword-parallel DVAFS multiplier is given in Chap. 4 in Sect. 4.5.1.2. For other building blocks, such as ALUs and multiplexers, other strategies are followed, depending on the relative importance of the building blocks.

Fig. 5.15 Three different supported modes in Envision. Each MAC in the processor can either compute on 1×1-16b DVAS input, or on 2×1-8b or 4×1-4b DVAFS inputs. (**a**) DVAS / DVAFS with $N = 1$. (**b**) DVAFS with N=2. (**c**) DVAFS with $N = 4$

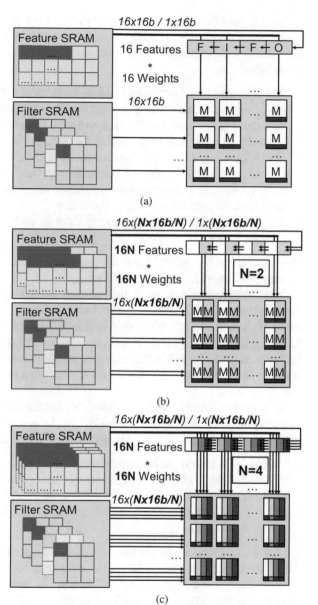

Making building blocks subword-parallel, three different strategies were followed depending on the impact a change would have on critical path delay and energy consumption. These three strategies were only used due to timing constraints.

1. Define custom building blocks in ASIP-designer's (Wu and Willems 2015) C-like PDG language. This strategy is followed for any adder-type operations, where the impact of this high-level language change is deemed minimal, as the changes necessary to create subword-parallel blocks are minimal.
2. Logic-doubling, where different hardware is used for different DVAFS modes is used for a small number of blocks that are not frequently used, but require major changes to be made subword-parallel. Examples of this are the ReLU and Max-Pool hardware blocks in the processor.
3. All multipliers and MACs are custom built in Verilog-RTL in order to have an efficient design.

Ideally, strategy 3—optimal hand-designed RTL subword-parallel blocks—should be used in all cases to minimize any overheads.

All DVAFS-compatible building blocks are controlled by a single programmable status register, similar to the strategy for DVAS-compatible building blocks. This status register is directly connected to all DVAFS-arithmetic for reconfiguration. This way, the chip can switch its operation between 1-16b DVAS modes and 4×1-4b, 2×1-8b DVAFS modes. The circuitry and software to reshuffle data operating in different modes from layer-to-layer is not included on-chip. This could be part of our future work.

Memory fetch- and operator-guarding, as discussed in detail in Sect. 5.2.3, cannot be performed on a per word basis in DVAFS, as this would incur large overheads. All guarding is performed at the default full-word level, not at the subword-level.

- Especially in memories, a single address could maximally contain 4 separate 4b words. In such implementation, memory-fetch guarding can only be performed at an address granularity and not at a subword-granularity.
- In the 2D-MAC array, inputs are only gated at a per-MAC granularity and not at a subword-granularity.

This effectively reduces the potential for energy gains in highly sparse models, as in the 4×4b case, all four words have to be zero before the system can properly guard memory-fetches and MAC-operations. Hence, in Envision V1, every zero-valued word can be guarded. In Envision V2, guarding is only possible if all subwords are zero-valued.

5.4.2   Physical Implementation

In order to build a successful DVAFS implementation, the same modifications as in DVAS are necessary: a physical design split into multiple power domains and

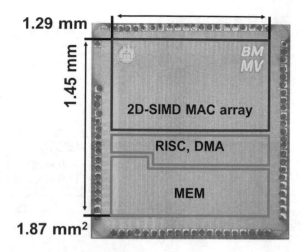

Fig. 5.16 Chip photograph and layout overview of the processor. The system totals 1.9 mm^2 in a 28 nm FD-SOI technology

1.29 mm

1.45 mm

BM
MV

2D-SIMD MAC array

RISC, DMA

MEM

1.87 mm^2

a multi-mode back-end optimization guaranteeing critical path scaling in DVA(F)S operating modes. As this design was made in 28 nm FD-SOI where transistor V_t can be modulated through modulating the *body-bias* voltage, several FD-SOI-specific steps have to be incorporated as well. In this case, every power domain is also a separate body-bias domain.

Hence, the full chip exists out of three power/body-bias domains: one containing all voltage-accuracy scalable *vas* DVAFS-compatible building blocks at a variable voltage, one containing non-accuracy scalable *nas* blocks such as the control logic and accuracy-scalable blocks *as* such as data transfer blocks, and one containing all memories. The design was optimized for 200 MHz typical operation in different simultaneously performed optimization modes.

Figure 5.16 is a photograph of the fabricated chip. The chip requires 1.9 mm^2 of active area in its 28 nm technology. Indicated in the figure is the placement of the different power/body-bias domains. Due to these splits, three supply voltages and 6 body-bias (2 per body bias domain) voltages can be independently optimized.

5.4.3 Measurement Results

A DVAFS-compatible processor has the potential to outperform a DVAS-scalable processor at reduced precision, as is clear from the discussion in Sect. 4.3 in Chap. 4. This section discusses the measurements on different operating modes in Envision and compares the DVAS-vs-DVAFS modes.

Fig. 5.17 Relative energy
consumption per word as a
function of computational
precision under DAS, DVAS,
and DVAFS

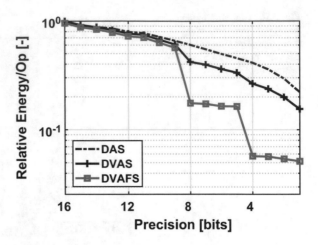

5.4.3.1   Performance Under DVA(F)S

A non-unrolled $5 \times 5 \times C$ CONV layer, similar to Listing 5.1, is executed on the
processor in different modes at a fixed real throughput of 76 GOPS. At full precision
and in DVAS, the operating frequency for the full system is then fixed to 204 MHz.
In DVAFS, the operating frequency is either 204 MHz at a computational precision
of 9-16b, 102 MHz at a precision ranging from 5 to 8b, and 51 MHz at a precision
ranging from 1 to 4b. On the arithmetic-building block level, little gains are to be
expected from DVAFS. However, in a system with large *nas* overheads such as
Envision, DVAFS reduces the overhead of *nas* blocks at low precision, allowing the
system to be more energy-efficient in these cases. In the DVAS-version of Envision,
for example, energy consumption is dominated by control and memory overheads
in the 1-4b precision region. These overheads can be reduced at least by a factor of
4 in a DVAFS-compatible architecture.

This is illustrated in the measurements shown in Fig. 5.17. Here, energy per
operation is plotted against computational precision at a fixed 76 GOPs throughput,
relative to the 16b full precision case in the same architecture. In DAS, energy/op is
only 5× lower than in the high precision case. This is due to the large clocking,
control, and memory overheads in this mode. Energy consumption is reduced
further up to a maximum of 8× at 1b compared to the full precision baseline due
to DVAS voltage scaling. In DVAFS, the remaining non-compute overheads can
be diminished at the same word-throughput by using the MACs in their 4-way
subword-parallel mode at the 1b-low-precision case. Here, relative energy per word
is 20× lower than the default 16b baseline.

A more detailed overview of the performance of Envision under DVAFS is shown
in Fig. 5.18. Here, the supply voltage and energy-efficiency under nominal body-
biasing operations is shown as a function of the required throughput in different
DVAFS modes. The shown $1 \times 16b$, $2 \times 8b$, $4 \times 4b$ non-sparse and 4×3-4b sparse
modes are all measured on the same $5 \times 5 \times C$ benchmark layer. The 4×3-4b

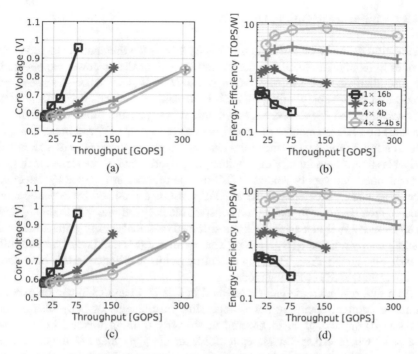

Fig. 5.18 Supply voltage and energy-efficiency as a function of average throughput in different DVAFS operating modes. (**a**) and (**b**) are with nominal body-bias, (**c**) and (**d**) are with optimal body-bias. Here optimality is defined as most energy-efficient at a given throughput

sparse mode is seen as representative for a low-precision always-on task. Weights are quantized to 3b, activations to 4b, while sparsity is 30% on the weights and 60% on the activations due to the quantization and ReLU operators. The nominal body-bias voltage in this case is non-zero at the nominal 1 V supply voltage. Rather, a symmetrical +/− 0.6 V body-bias voltage is chosen in order to allow a nominal operating frequency of 204 MHz at nominal supply. Although the design was optimized to be able to run at 204 MHz at nominal supply and zero body-bias, non-zero nominal biasing was necessary in the measurement setup. Up to this time, it is still unclear what caused the slower operation of this chip in measurement. From these measurements, it is clear that energy-efficiency can be pushed up from 0.33 TOPS/W in 16b full precision at a core voltage of 1.05 V up to 4 TOPS/W 0.66 V in 4 × 4b mode and 8 TOPS/W at 0.62 V in a sparse 4×3-4b operating mode. In the low-precision case, throughput can be pushed upwards to 300 GOPS if the system is operated at 204 MHz. In this case, energy-efficiency drops to 2.5 TOPS/W in the default 4 × 4b mode and to 6 TOPS/W in the sparse low-precision mode.

5.4.3.2 Influence of Optimal Body-Biasing

Modulating the body-bias of a circuit directly affects the threshold voltage V_t of all devices in its domain. Reducing the V_t increases the leakage power exponentially, but also increases the overdrive voltage $VDD - V_t$ and hence the speed of a circuit at a fixed supply voltage VDD. Increasing V_t reduces the leakage power exponentially and does not change dynamic power consumption considerably. At the same time, it also reduces the overdrive voltage $VDD - V_t$ and hence the speed of the circuit. Depending on the initial leakage-vs-active power ratio of a circuit, the body-bias thus allows increasing efficiency at a fixed speed. If the same clock speed is maintained, the supply voltage VDD can be reduced at a lower V_t, effectively increasing energy-efficiency if the system power is dominated by dynamic power consumption. In order to maintain a constant clock speed, the VDD should hence be increased if V_t is increased. This can still lead to global energy gains if the power is significantly dominated by leakage. The optimal VDD and body-bias controlled V_t will vary across $P_{leak}/P_{dynamic}$ operating points and hence in this case across the different DVAFS operating modes.

As a rule-of-thumb the gains described in Eqs. (5.1) and (5.2) can be expected around a nominal operating point, depending on the type of body-bias (reverse or forward) applied. If V_t is lowered nonlinearly through forward body-biasing to a point where leakage increases tenfold, the supply voltage at the same clock frequency can be lowered by more than $\sqrt{2}$ or a decrease of 30%. The global power reduction in this case hence depends on the ratio of leakage-vs-dynamic power consumption. If initial leakage is around 1%, the strategy makes sense and leads to a global reduction.

$$P_{total} = P_{dyn}/2 + P_{leak} \times 10 \qquad (5.1)$$

If leakage is close to 10%, global power will increase and the strategy should be avoided. At such balance in leakage-vs-dynamic power, the reverse strategy could reduce power consumption. If V_t is increased through reverse body-biasing, leakage could decrease tenfold, while dynamic power consumption at the same clock frequency increases by about a factor of 1.7, if the supply voltage is increased by about 30%. Power consumption can then be estimated as follows:

$$P_{total} = P_{dyn} * 1.7 + P_{leak}/10 \qquad (5.2)$$

Equations (5.1) and (5.2) are visualized in Fig. 5.19. From this graph it is clear that forward body-biasing only makes sense when the percentage of leakage power is in the 1–5% range. Gains can be expected from reverse body-biasing when the contribution of dynamic power is below 50%. In general, body-biasing should only be applied in extreme modes: either when leakage is dominant, or when dynamic power is dominant.

This kind of dynamic power scaling is specifically applicable to Envision V2, as its different operating modes have different $P_{leak}/P_{dynamic}$ ratios. The system

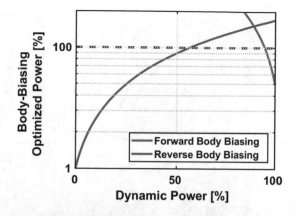

Fig. 5.19 First order expected gains using body-biasing in FDSOI, starting from a nominal operating point. The expected gains depend strongly on the ratio between dynamic and leakage power

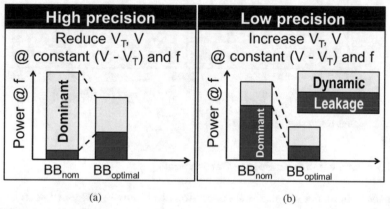

(a)   (b)

Fig. 5.20 Body-biasing rationale in Envision. (**a**) BB in high precision modes. (**b**) BB in low-precision modes

can hence be differently optimized by different body-bias and supply voltage combinations in every mode. In Fig. 5.20a, power consumption is dominated by dynamic switching energy. This opens opportunities to reduce energy consumption through body-biasing. At low precision, as in Fig. 5.20b energy consumption can be leakage dominated. An optimized body-biasing strategy reduces leakage at a slightly higher dynamic cost.

Full measurement results using optimal body-bias settings are given in Fig. 5.18c, d. In high precision modes, body-biasing allows reducing the supply voltage and hence increases energy-efficiency. In the low-precision modes, energy-efficiency is improved by lowering the leakage power in these modes. Ultimately this leads to a peak energy-efficiency measured in Envision of 10 TOPS/W at a throughput of 76 GOPS in the sparse 4×3-4b operating mode.

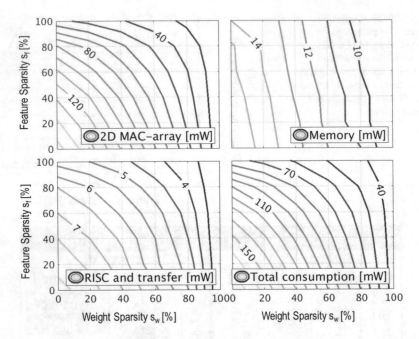

Fig. 5.21 Sparse processing in Envision

5.4.3.3   Performance on Sparse Datastreams

The general influence of sparse data operation on effective energy-efficiency is given in Fig. 5.21. Here, total power consumption at 152 GOPS throughput and 200 MHz in the 2 × 8b mode is split into three separately measurable parts: the 2D-MAC array, all memory and the control, and data transfer parts. Total power consumption is indicated as well. In these plots, both the sparsity in the feature input streams and in the weight input streams is modulated from 0 to 100%. Due to the FIFO-based dataflow on the feature stream, power consumption in the memory part will be highly dominated by the weight- and program-memory buffers. Hence, the influence of feature sparsity on memory consumption is limited, while it remains high on the power consumed in the 2D-MAC.

5.4.3.4   Performance on Benchmarks

The combined effect of the proposed techniques on the energy consumption of Envision is further measured on the CONV layers of three benchmarks: AlexNet(Krizhevsky et al. 2012), Lenet-5 (LeCun et al. 1998), and the VGG-16 net that is also used in the face recognition hierarchy as the final stage. All measurements are shown in Fig. 5.22 and are performed at the indicated frequency $f = 204$ MHz/N, where N is the DVAFS mode, equal to the level of subword

Layer	Weight bits	Input bits	Parallel Mode / frequency [MHz]	Weight sparsity (%)	Input sparsity (%)	Weight BW Reduction	Input BW. Reduction	mem / ctrl / 2D / BB Voltage (V)	MMACs/ frame	Power (mW)	Efficiency (TOPS/W)
Precise	16	16	1x16b / 200	0	0	1.00x	1.00x	1 / 1 / 1 / +- 1.2	-	**290**	0.3
VGG l1	5	4	2x8b / 100	5	10	1.00x	1.03x	1 / .65 / .65 / +- 0.8	87	25	2.1
VGG l2	5	6	2x8b / 100	25	51	1.23x	1.81x	1 / .65 / .65 / +- 0.8	1850	31	1.6
VGG l3	5	6	2x8b / 100	33	30	1.36x	1.31x	1 / .65 / .65 / +- 0.8	924	35	1.5
VGG l4	5	6	2x8b / 100	40	30	1.51x	1.31x	1 / .65 / .65 / +- 0.8	1850	34	1.5
VGG l5	5	6	2x8b / 100	75	52	3.20x	1.84x	1 / .65 / .65 / +- 0.8	925	25	2.1
VGG l6	5	6	2x8b / 100	56	47	2.00x	1.69x	1 / .65 / .65 / +- 0.8	1850	27	1.9
VGG l7	5	6	2x8b / 100	55	47	1.95x	1.69x	1 / .65 / .65 / +- 0.8	1850	27	1.9
VGG l8	5	6	2x8b / 100	64	63	2.37x	2.31x	1 / .65 / .65 / +- 0.8	925	22	2.4
VGG l9	5	6	2x8b / 100	74	63	3.10x	2.31x	1 / .65 / .65 / +- 0.8	1850	21	2.4
VGG l10	5	6	2x8b / 100	73	61	3.00x	2.21x	1 / .65 / .65 / +- 0.8	1850	22	2.4
VGG l11	5	6	2x8b / 100	38	82	1.47x	4.12x	1 / .65 / .65 / +- 0.8	462	20	2.6
VGG l12	5	6	2x8b / 100	36	82	1.42x	4.12x	1 / .65 / .65 / +- 0.8	462	20	2.6
VGG l13	5	6	2x8b / 100	68	76	2.61x	3.31x	1 / .65 / .65 / +- 0.8	462	19	2.8
Total	-	-	-	-	-	-	-	-	15346	**26**	2
AlexNet l1	7	4	2x8b / 100	21	29	1.17x	1.30x	1 / .65 / .65 / +- .8	104	37	2.7
AlexNet l2	7	7	2x8b / 100	19	89	1.15x	5.80x	1 / .65 / .65 / +- .8	224	20	3.8
AlexNet l3	8	9	1x16b / 200	11	82	1.05x	4.10x	1 / 1 / .85 / +- 1.2	150	52	1
AlexNet l4	9	8	1x16b / 200	4	72	1.00x	2.90x	1 / 1 / .85 / +- 1.2	112	58	0.9
AlexNet l5	9	8	1x16b / 200	4	72	1.00x	2.90x	1 / 1 / .85 / +- 1.2	75	62	0.8
Total	-	-	-	-	-	-	-	-	666	**44**	1.8
LeNet-5 l1	3	1	4x4b / 50	35	87	1.40x	5.20x	1 / .65 / .65 / +- .4	0.3	**5.6**	13.6
LeNet-5 l2	4	6	2x8b / 100	26	55	1.25x	1.90x	1 / .65 / .65 / +- .8	1.6	29	2.6
Total									1.9	**25**	3

Fig. 5.22 Envision's performance on the CONV layers of LeNet-5, AlexNet, and VGG-16

parallelism. More specifically, Fig. 5.22 shows the effect of computational precision and sparsity on supply voltage, optimal body-biasing, IO compression, and power consumption. At full 16b precision, power consumption is around 290 mW at 200 MHz, $+/-1.2$ V body-biasing, and 1 V supply on all power domains. This power consumption drops to below 6 mW at 50 MHz in the 4×1-3b DVAFS mode of layer 1 in LeNet-5. Here, body-biasing is reduced to $+/-.4$ V, reducing the leakage power significantly. Power is further minimized due to sparsity guarding and advanced voltage scaling on all but the memory power domains.

A more detailed example for AlexNet layer 2 is given in Fig. 5.23. It shows the effect of our precision scaling (C) and sparsity exploitation techniques (B) on this specific layer on Envision V2. Without any optimizations on this layer, it operates in the $N = 1$ mode, where every feature and weight is represented using 16b. Sparsity guarding is shut down as well. In this mode, the chip consumes 290 mW at a throughput of 76 GOPS. However, that specific layer can operate at 7b if it is permitted to lose 1% of relative benchmark accuracy. Hence, the chip can operate in mode $N = 2$ processing 2 words per operating units per cycle. If throughput is kept constant at 76 GOPS, this leads to a 5.4× decrease in energy consumption, because of voltage, activity, and frequency scaling. On top of that, AlexNet layer 2 is exceptionally sparse, up to 90% in its feature stream. This leads to an additional 2.5× reduction in energy consumption, totaling 15× compared to the non-optimized baseline. In other layers, the gains compared to 16b non-sparse operation are more limited, but still significant.

On top of that, Envision V2 is benchmarked in a face recognition hierarchy, similar to the one discussed in Chap. 2. Figure 5.24 compares the performance of Envision on the CONV layers of separate steps of such a face recognition

Fig. 5.23 The gains of exploiting sparsity (B) and reduced precision (C) operation compared to a dataflow optimized baseline (A)

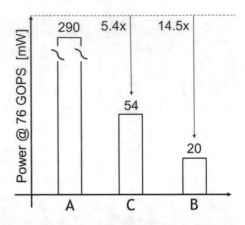

ConvNet	Input	Network topology [+]	# Output Classes	Feature [bits]	CONV [bits]	FC [bits]	CONV Size [B]	FC Size [B]	CONV Energy [uJ]	Acc. [%]
Face-vs-background	32x32x3	C3x32_C3x48_C3x64_F64_F64_F2	2	2 to 3	4	2	22k	34k	3	94
Owner-vs-all	32x32x3	C3x48_C3x64_C3x96_F96_F64_F3	3	3 to 4	4	3	42k	77k	6	96
10 Faces	32x32x3	C3x64_C3x128_C3x128_FC128_FC64_FC12	12	4 to 6	4	4	112k	135k	35	95
100 Faces	224x224x3	C5x128_C3x256_C3x256_C3x256_FC512_FC256_FC104	104	4 to 6	5	6	750k	3.3M	552	94
VGG-16	224x224x3	VGG-16	5760	5 to 6	6	7	15M	124M	23100	90=95

Fig. 5.24 Envision V2 CONVL energy consumption and achieved accuracy on different sub-tasks in a large-scale face recognition hierarchy, similar to the one discussed in Chap. 2

hierarchy. The different steps of the hierarchy achieve high precision on their respective data sets at different bit widths. Envision can then specifically minimize energy consumption for these cases by using its DVAFS capabilities. The smallest networks "face-vs-background" and "owner-vs-all" are sufficiently small so they fit completely on chip.

5.4.3.5   Comparison with the State of the Art

Envision V2 is compared with other SotA accelerators in Figs. 5.25 and 5.26.

Envision scales efficiency on the AlexNet convolutional layers between 0.8 and 3.8 TOPS/W depending on the layer sparsity, the layer topology, and its required precision. This should be compared to 0.16 TOPS/W (Chen et al. 2016), 0.56–1.4 TOPS/W, and 0.2–1.1 TOPS/W in the 40 nm DVAS-compatible version of Envision (Moons and Verhelst 2016). Efficiency is 2 TOPS/W average for the

	[Cav15a] GLSVLSI '15	[Che16] ISSCC '16	[Moo16] VLSI '16	**This work** **N = 1, 2 or 4**
Technology	65nm CMOS	65nm LP CMOS	40nm LP CMOS	**28nm UTBB FD-SOI**
Nominal Frequency [MHz]	500	200	200	**200**
Supply @ f_{nom} [V]	1.2	1	1.1	**1**
Peak performance [GOPS]	Fixed 196	Fixed 67	Fixed 102	**Dynamic N x 102**
Active Area [mm²]	1.31	12.25	2.4	**1.87**
# of MACs	-	168	256	**Dynamic N x 256**
Gate Count [NAND-2]	0.9M	1.852M	1.6M	**1.95M**
On-Chip SRAM [kB]	43	184.5	144	**144**
# layers, # filters [-]	All	All	All	**All**
Filter sizes [-]	<7x7	1-1024	All	**All**
Precision [bits]	Fixed 12	Fixed 16	Dynamic 1-16	**Dynamic N x 1-16 / N**
AlexNet Conv-layers [mW]	-	278 @ 34.7fps	55-95, 76 avg. @ 47fps	**20-62, 44 avg. @ 47fps**
VGG Conv-layers [mW]	-	-	-	**19-35, 26 avg. @ 1.67fps**
Dynamic power range @ GOPS$_{nom}$ [mW]	510 (1x) @ 145 GOPS	235-332 (1.5x) @ 46 GOPS	35-300 (8.5x) @ 80 GOPS	**7.5-300 (40x)** **@ 76 GOPS**
Min. efficiency [TOPS/W]	0.44	0.17	0.27	**0.26**
Max. efficiency [TOPS/W]	0.8	0.25	2.6	**10**

Fig. 5.25 Comparison of Envision V1 and V2 to previous references in the field

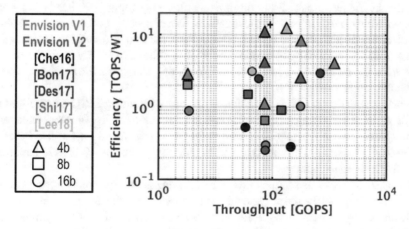

Fig. 5.26 Visual representation of SotA CNN-accelerators in the throughput-efficiency-computational precision plane. Envision V2 outperforms V1 up to 4× in terms of maximum energy-efficiency. + indicates the measurement is performed on a sparse datastream. All other measurements are on non-sparse streams

CONV layers of VGG-16 and up to 10 TOPS/W peak on a sparse $4 \times 4b$ layer. Furthermore, unlike any other reference except Envision v1, this DVAFS version of Envision is the only one that can significantly scale power consumption depending on computational precision and sparsity. The chip allows a global 40× trade-off from a full precision baseline down to a low-precision sparse mode. This further illustrates Envisions capability to minimize energy consumption for any CNN, depending on its system-level requirements. The chip hereby enables always-on hierarchical recognition applications.

5.4.4   Envision V2 Overview

A precision-scalable processor for CNNs with operator- and memory-guarding was fabricated in a 28 nm FD-SOI technology. It has a total active area of 2 mm^2 and runs at a nominal frequency of 204 MHz at 1.05 V and nominal $+/-0.6$ V body-biasing.

First, the processor is fully C-programmable and uses a reconfigurable 256-1024 2D-MAC array architecture as an efficient convolution baseline. This 2D-architecture allows inherent 16× reuse of filter weights and feature map inputs. A FIFO further reduces internal memory bandwidth up to 27.8×.

Second, the processor minimizes energy by modulating precision and supply voltage in the MAC array from layer-to-layer using the DVAFS principle. This requires extra rounding circuitry, subword-parallel DVAFS-compatible building blocks, a split into three power domains, and an advanced back-end place and route optimization. Compared to DVAS-only, energy-efficiency is increased more in DVAFS, as the technique allows minimizing all non-compute overheads that would otherwise become dominant in those modes. When scaling down from $1 \times 16b$- to $2 \times 8b$ or $4 \times 4b$ at 76 GOPS constant throughput, the scalable supply voltage can go down from 1 V nominally at nominal body-bias voltage to 0.65 V at different boy-bias voltages, respectively. In AlexNet layer 2, this leads to a 5.4× gain compared to the full precision baseline.

Finally, sparsity in CNNs is exploited through preventing SRAM-banks and the 2D array from switching and by compressing data up to 5.8×. In AlexNet layer 2, sparsity leads to a 2.7× gain compared to the precision and voltage scaled baseline, while only 32 sparsity flags are required to potentially prevent 32 SRAM banks and 256 MACs from switching.

Because of these techniques this chip minimizes energy consumption for any CNN, showing 4–10× improvement over the state of the art in terms of energy/frame using 4b operations in Envision V1 and a minimum of 20% higher than a 2017 alternative (Shin et al. 2017). Envision V2 can reduce its energy by 15× compared to non-sparse full precision operation in the same architecture on AlexNet layer 2. The maximum theoretical gain factor at constant throughput is up to 40×, a fact that can be well exploited in the hierarchical systems of Chap. 2. The chip hereby enables low-power, high-performance applications of computer vision for battery powered devices.

5.5 Conclusion

This chapter discussed the design and measurements of two generations of energy-scalable CNN processors. These processors offer (a) an energy-efficient baseline in a parallel 2D-MAC array architecture. They can minimize energy consumption depending on (b) the sparsity of a CNN-application and (c) its computational precision requirements.

(a) The 2D-MAC array with input FIFO offers a high throughput increase, while minimizing bandwidth to the on-chip memory. Only 32 words need to be fetched per cycle to generate 256 outputs per cycle. The nature of the 2D-system allows rounding and guarding inputs to the array efficiently, with limited energy overhead.

(b) Finally, Envision exploits the sparsity of weight and feature input streams to minimize energy consumption further. A separate on-chip memory contains flags with sparsity information that are used to effectively gate memory-fetches and MAC-operations in the 2D-array. In Envision, this can lead to an additional 2–$2.7\times$ increase in energy-efficiency, if the input datastreams are 50–90% sparse.

(c) Both versions of the architecture can modulate the used computational precision dynamically on a layer-per-layer basis and scale energy consumption accordingly. In V1 this is done through dynamic-voltage-accuracy-scaling (DVAS), which modulates supply voltage on the 2D-MAC array at lower computational precision. In V2, this is expanded upon by making the architecture subword-parallel and hence dynamic-voltage-accuracy-frequency-scalable (DVAFS) compatible, which increases the energy-accuracy trade-off considerably. Energy-efficiency is specifically reduced at low precision, as the DVAFS techniques also reduces the contribution of all non-compute overheads, such as control, data transfers, and SRAM energy. Through DVA(F)S, operating on 1b operands is $8\times$ $(20\times)$ more efficient than the 16b full precision baseline. Depending on the application, this may not lead to a loss in system-level accuracy.

Figure 5.26 shows Envision V1 and V2 together with other recent references in the throughput vs energy-efficiency space. Envision V2 shows one of the highest achieved energy-efficiencies in terms of 4b ops, outperforming V1 4–$10\times$ at iso-throughput and the closest SotA by 20% at a lower throughput, albeit for a sparse dataflow. Lee et al. (2018) is a 1–16 bit weight implementation for CNNs. It shows the highest 4b energy-efficiency at a high effective throughput, up to 11.6 TOPS/W. The same chip is also very competitive at full 16b precision, showcasing 3.06 TOPS/W at 16b and 43 GOPS.

References

Albericio J, Judd P, Jerger N, Aamodt T, Hetherington T, Moshovos A (2016) Cnvlutin: ineffectual-neuron-free deep neural network computing. In: International symposium on computer architecture (ISCA)

Cavigelli L, Gschwend D, Mayer C, Willi S, Muheim B, Benini L (2015a) Origami: a convolutional network accelerator. In: Proceedings of the 25th edition on Great Lakes symposium on VLSI, pp 199–204

Cavigelli L, Magno M, Benini L (2015b) Accelerating real-time embedded scene labeling with convolutional networks. In: Proceedings of the 52nd annual design automation conference

Chen Y, Luo T, Liu S, Zhang S, He L, Wang J, Li L, Chen T, Xu Z, Sun N, et al (2014) DaDianNao: a machine-learning supercomputer. In: Proceedings of the 47th Annual IEEE/ACM international symposium on microarchitecture, pp 609–622

Chen YH, Krishna T, Emer J, Sze V (2016) Eyeriss: An energy-efficient reconfigurable accelerator for deep convolutional neural networks. ISSCC Dig of Tech papers, pp 262–263

Chippa VK, Venkataramani S, Chakradhar ST, Roy K, Raghunathan A (2013) Approximate computing: an integrated hardware approach. In: Proceedings of the Asilomar conference on signals, systems and computers, pp 111–117

Du Z, Fasthuber R, Chen T, Ienne P, Li L, Luo T, Feng X, Chen Y, Temam O (2015) ShiDianNao: shifting vision processing closer to the sensor. In: International symposium on computer architecture (ISCA), pp 92–104

Han S, Liu X, Mao H, Pu J, Pedram A, Horowitz MA, Dally WJ (2016) EIE: efficient inference engine on compressed deep neural network. In: International symposium on computer architecture (ISCA)

Huffman DA, et al (1952) A method for the construction of minimum-redundancy codes. Proc IRE 40(9):1098–1101

Knag P, Chester L, Zhang Z (2016) A 1.40 mm^2 141mW 898 GOPS sparse neuromorphic processor in 40 nm CMOS. In: Proceedings of the IEEE symposium on VLSI circuits, pp 180–181

Krizhevsky A, Sutskever I, Hinton GE (2012) Imagenet classification with deep convolutional neural networks. In: Proceedings of advances in neural information processing systems, pp 1097–1105

LeCun Y, Bottou L, Bengio Y, Haffner P (1998) Gradient-based learning applied to document recognition. Proc IEEE 86(11):2278–2234

Lee J, Kim C, Kang S, Shin D, Kim S, Yoo HY (2018) UNPU: a 50.6 tops/w unified deep neural network accelerator with 1b-to-16b fully-variable weight bit-precision. In: International solid-state circuits conference (ISSCC)

Moons B, Verhelst M (2016) A 0.3–2.6 TOPS/W precision-scalable processor for real-time large-scale convNets. In: Proceedings of the IEEE symposium on VLSI Circuits, pp 178–179

Moons B, Uytterhoeven R, Dehaene W, Verhelst M (2017) Envision: a 0.26-to-10 TOPS/W subword-parallel dynamic-voltage-accuracy-frequency-scalable convolutional neural network processor in 28nm FDSOI. In: International solid-state circuits conference (ISSCC)

Motamedi M, Gysel P, Akella V, Ghiasi S (2016) Design space exploration of FPGA-based deep convolutional neural networks. In: Proceedings of the 21st Asia and South Pacific design automation conference (ASP-DAC), pp 575–580

Rahman A, Lee J, Choi K (2016) Efficient FPGA acceleration of convolutional neural networks using logical-3D compute array. In: Proceedings of the design, automation & test in Europe conference & exhibition (DATE), pp 1393–1398

Reagen B, Whatmough P, Adolf R, Rama S, Lee H, Lee SK, Hernandez-Lobato JM, Wei GY, Brooks D (2016) Minerva: enabling low-power, highly-accurate deep neural network accelerators. In: Proceedings of the ACM/IEEE 43rd annual international symposium on computer architecture (ISCA)

Shin D, Lee J, Lee J, Yoo HJ (2017) 14.2 DNPU: an 8.1 TOPS/W reconfigurable CNN-RNN processor for general-purpose deep neural networks. In: 2017 IEEE international solid-state circuits conference (ISSCC). IEEE, New York, pp 240–241

Suda N, Chandra V, Dasika G, Mohanty A, Ma Y, Vrudhula S, Seo Js, Cao Y (2016) Throughput-optimized openCL-based FPGA accelerator for large-scale convolutional neural networks. In: Proceedings of the 2016 ACM/SIGDA international symposium on field-programmable gate arrays, pp 16–25

Vanhoucke V, Senior A, Mao MZ (2011) Improving the speed of neural networks on CPUs. In: Deep learning and unsupervised feature learning workshop at advances in neural information processing systems

Wu B, Willems M (2015) Rapid architectural exploration in designing application-specific processors. In: ASIP designer whitepaper

Chapter 6
BINAREYE: Digital and Mixed-Signal Always-On Binary Neural Network Processing

6.1 Binary Neural Networks

6.1.1 Introduction

As discussed in Chaps. 1 and 3, Convolutional Neural Networks (CNN) are state-of-the-art deep learning algorithms that are too costly to embed on wearable, battery constrained devices. One of the ways to reduce both their computational and memory footprint is through reducing the precision of the used weights and activations. Reduced precision operators can reduce the cost of arithmetic operations and the relative cost of the control logic, especially if techniques like DVAFS (Chap. 4) are used. Lowering the precision of weights and activations in QNN's also potentially reduces a system's memory footprint: fewer bits per stored word means lower on-chip memory requirements and less communication with main memory. Chapter 3 shows how to choose the used number of bits optimally for a number of benchmarks. BinaryNets (Hubara et al. 2016)—networks with all weights and activations constrained to $+/-1$—are the ultimate form of QNN and hence promise significant energy savings for ultra-low power applications.

In BinaryNets, a digital $1b'0$ represents the number -1, while a digital $1b'1$ represents the number $+1$. At training time, floating-point weights and activations are used for computing all parameter gradients, as discussed in Sect. 3.4 in Chap. 3. During the forward pass, BinaryNets reduce memory size and accesses by $32\times$ compared to a floating-point baseline and replace almost all internal operations by bit-wise operations. Of course, due to the low *modeling capacity* of 1b weights and activations, a BinaryNet will have to be wider and deeper than its high-precision counterparts. This trade-off is investigated in detail in Chap. 3.

© Springer Nature Switzerland AG 2019
B. Moons et al., *Embedded Deep Learning*,
https://doi.org/10.1007/978-3-319-99223-5_6

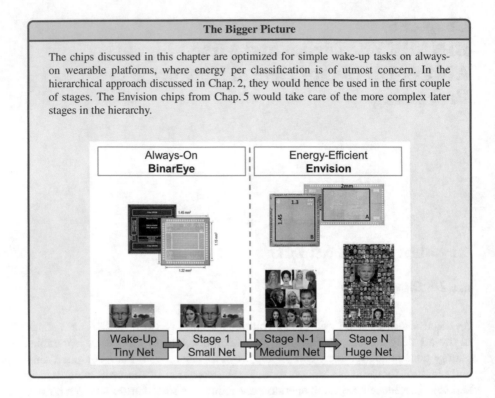

The Bigger Picture

The chips discussed in this chapter are optimized for simple wake-up tasks on always-on wearable platforms, where energy per classification is of utmost concern. In the hierarchical approach discussed in Chap. 2, they would hence be used in the first couple of stages. The Envision chips from Chap. 5 would take care of the more complex later stages in the hierarchy.

6.1.2 Binary Neural Network Layers

As in any deep neural network, a BinaryNet consists of a sequence of CONV-, FC-, ACT-, Max-pooling, and batch normalization layers. However, compared to high-precision neural networks, several of the operators used in those layers can be simplified significantly.

As discussed in Chap. 1, a **Convolutional Layer** (CONVL) transforms input feature maps (I) into output feature maps (O), each containing multiple units. Each unit in an output feature map ($M \times M \times F$) is connected to local patches of units ($K \times K \times C$) in the input feature maps through a filter $W[F]$ ($K \times K \times C \times 1$) in a filter bank W ($K \times K \times C \times F$) existing out of a set of machine-learned weights and a bias (B) per output feature map. A formal mathematical description is given in the following equation:

$$O[f][x][y] = \sum_{c=0}^{C} \sum_{i=0}^{K} \sum_{j=0}^{K} I[c][Sx + i][Sy + j] \times W[f][c][i][j] \\ + B[f]$$

(6.1)

where S is a stride and x, y, f are bounded by: $x, y \in [0, \ldots, M[$ and $f \in [0, \ldots, F[$. As all internal weights and activations are either $+/-1$, a multiplication in this domain is only a simple XNOR operation. The previous equation can hence be simplified as follows:

$$O[f][x][y] = \text{POPCOUNT}^{C,K,K} \text{XNOR}(I[c][Sx + i][Sy + j], W[f][c][i][j])$$
$$+ B[f]$$

(6.2)

Here, the bias $B[f]$ and the output $O[f][x][y]$ are signed integers with a limited range that is known at train time. The POPCOUNT operator is the same summation over the same dimensions as in Eq. (6.1). However in the binary case, it can be simplified to a simple counter, rather than a general addition.

Similarly in a **fully connected layer** (FC), the standard neuron operation is

$$O[z] = \sum_{m=0}^{M} W[z, m] \times I[m] + B[z]$$

(6.3)

which can be replaced by:

$$O[z] = \text{POPCOUNT}^{M} \text{XNOR}(W[z, m], I[m]) + B[z]$$

(6.4)

This neuron operator hence uses exactly the same operators as the CONV layer. The only difference is in the weight reuse: all weights in an FC layer are used only once.

The result of the local sum computed in this filter bank is then passed through a **nonlinearity layer**. In a SotA neural network, this is typically a Rectified Linear Unit (ReLU), using the nonlinear activation function $a(u) = max(0, u)$, where u is a feature map unit. In BinaryNets, this activation is simplified to a hard tangent hyperbolic operation, which simplifies to the sign function in the binary forward pass:

$$a(u) = sign(u)$$

(6.5)

where u is again a feature map unit. Note that a sign function is extremely easy to implement in digital electronics. It merely requires reading the MSB of u.

Max-pooling layers, shown in Fig. 6.1, compute and output only the maximum of a local patch (typically 2×2 or 3×3) of output units in a feature map. They thereby reduce the dimension of the feature representation and create an invariance to small shifts and distortions in the inputs. On a binary feature map, where all values are constrained to $+/-1$ after activation, a max-pooling operation simply comes down to checking if any of the values in the patch are 1, or equal to digital

Fig. 6.1 Max-pooling visualized

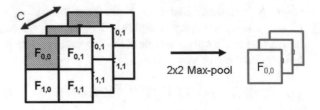

$1'b1$. Max-pooling can be implemented as an N-way digital *OR* in digital logic. This is a very cheap operation on 2×2 or 3×3 patches.

Finally, BinaryNets require **batch normalization** (Ioffe and Szegedy 2015), as first discussed in Chap. 1 to prevent overfitting and ensure good and fast convergence during training. Batch normalization normalizes the inputs to subsequent layer to reduce the otherwise large *internal covariate shift*. This is done according to the statistics of the current batch. The output of a batch normalization layer y is then:

$$y = \frac{x - \mu}{\sqrt{\sigma^2 + \epsilon}} \gamma + \beta \qquad (6.6)$$

where x and y are the input and output of the batch normalization layer, μ is the mean of the current batch, σ is the standard deviation of the current batch, and ϵ is a factor ensuring numerical stability. γ and β are trainable shifting factors. None of the operators given here are inherently binary, which means that there is a potential need for internal floating-point operations if the batch normalization operation is to be implemented. Fortunately, due to the use of the sign activation function in BinaryNets, batch normalization does not have to be computed at run-time if the batch normalization layer follows a convolutional layer. The effect of batch normalization can be embedded into the previous CONVL at run-time. This process is described below.

First, the batch normalization process itself can be simplified to:

$$y = (xK) + (\beta - \mu K)$$
$$\text{with } K = \frac{\gamma}{\sqrt{\sigma^2 + \epsilon}} \qquad (6.7)$$

Tips and Tricks
Batch normalization does not require floating-point computation in BinaryNets. The sequence "NNL—Batch Normalization—sign/hardtanh activation" can be replaced with an equivalent "NNL_{eq}—sign/hardtanh activation" in the forward pass. This equivalent form can be computed offline after training and before the model is deployed.

> **Tips and Tricks**
>
> When using batch normalization on binary neural networks, or quantized neural networks in general, beware not to use the framework—standard settings provided by the likes of keras and tensorflow. Quantized neural networks require much lower momentum than their floating-point counterparts. An $\epsilon = 0.0001$ and momentum$= 0.1$ instead of the default 0.99 are experimentally verified to be good settings for quantized training.

Here, the second term is a biasing term that can be computed offline at train time and can be incorporated into the CONV or FC layer at test time. The full bias is then equal to $\beta_{EQ} = \beta_{\text{CONV}} + \beta_{BN}$. If batch normalization comes after a fully binary CONVL or FCL and if it is followed by a hardtanh/sign activation, its operations can be incorporated into the weights and bias of the previous layer. The output of a sequence of NNL—Batch Norm—Activation can be replaced as shown below. Here, $a(y)$ is the hardtanh activation function performed on y, the output of the batch normalization layer that takes x as an input. x is an output feature generated by a neural network layer with weights W and bias b.

$$
\begin{aligned}
A(y) &= sign(y) \\
&= sign(xK + (\beta - \mu K)) \\
&= sign(W \cdot a_{i-1} K + (\beta - (\mu - b)K)) \\
&= sign\left(|K|((sign(K)W) \cdot y_{i-1} + \left(\frac{\beta}{|K|} - (\mu - b)sign(K)\right)\right) \quad (6.8) \\
&= sign\left(((sign(K)W) \cdot y_{i-1} + \left(\frac{\beta}{|K|} - (\mu - b)sign(K)\right)\right) \\
&= sign(W_{EQ} \cdot y_{i-1} + \beta_{EQ})
\end{aligned}
$$

Hence, batch normalization is embedded into the NNL's bias and weights by computing equivalent sets $W_{EQ} = sign(K) \times W$ and $\beta_{EQ} = \frac{\beta_{BN}}{|K|} + (b - \mu)sign(K)$ for every layer. This trick is only possible for BinaryNets and not for general quantized neural networks, as only they use the sign/hardtanh activation function.

Apart from vanilla BinaryNets, several expansions on the same ideas have been developed with better results on large-scale benchmarks such as IMAGENET (Russakovsky et al. 2015). In Binary Weight networks (Courbariaux et al. 2015), only the weights are quantized to 1b. These do not seem to outperform fully quantized BinaryNets in the energy-accuracy space due to the large intermediate feature maps (Moons et al. 2017a). The BinaryNet training setup can also be generalized to TernaryNets (Zhu et al. 2016), nonlinear trained-clustered quantization (Han et al. 2016) and ultimately quantized neural networks (Moons et al. 2017a). BinaryNets themselves have been improved upon by moving towards XNOR-Nets (Rastegari et al. 2016) with much higher accuracy on ImageNet, for example. In XNOR-Nets, the core CONV layers are performed with binary precision as well, but these are followed by an affine transformation. However, the algorithm requires on the fly computation of these scaling factors based on the floating-point values of the input feature maps. All these are performed with floating-point precision. The XNOR-Net requires storing intermediate feature maps in high precision as well, which has high cost as discussed in Moons et al. (2017a), Sze et al. (2017), and Chap. 3. It is not yet clear how plain BinaryNets compared to XNOR-Nets in the energy-accuracy space, but due to the added floating-point complexity, seem less suitable for always-on wake-up applications.

6.2   Binary Neural Network Applications

To date, plain BinaryNets are unproven on complex large-scale image recognition tasks such as the ImageNet (Russakovsky et al. 2015) challenge. However, they do achieve high accuracy within reasonable range of the SotA on simpler tasks such as handwritten digit recognition (LeCun and Cortes 1998) (up to 98.5%), 10-class object recognition (Krizhevsky and Hinton 2009) (up to 88% on FPGA (Zhao et al. 2017)), and some of our own custom designed tasks such as face detection, owner recognition, 10-face recognition, and angle recognition (Moons et al. 2018, 2017c). This scope of applications, combined with their potential for low-energy operation, makes them suitable for use on battery constrained devices where an always-on gesture, face, owner, and angle detector can be used as a wake-up sensor for its screen or application processor. A specific future application, currently under industrial development, is an always-on camera on the front of a smartphone. This camera is equipped with recognition hardware that allows tracking a user's face and its angle, allowing autorotate functionality without relying solely on accelerometer data. Future generations of such devices could use BinaryNet ASICs.

Figure 6.2 situates the use of BinaryNets in the application and associated platform space. The y-axis in this space is energy efficiency: the number of classifications a system can perform per Joule of consumed energy. The x-axis is more abstract. It represents all quantities associated with *performance* in a broad sense: benchmark accuracy and throughput. In order to achieve this, the bandwidth and processing power a network requires typically increases as well. In this space, three regions are to be differentiated, each with their own associated energy efficiency, which is equivalent to a power budget in real-time processing. The highest performance is achieved in HPC systems such as GPU's and TPU's

---

The Bigger Picture

BinaryNets are trained to use 1b operators only. They are hence a special case of the quantized neural networks discussed in Chap. 3.

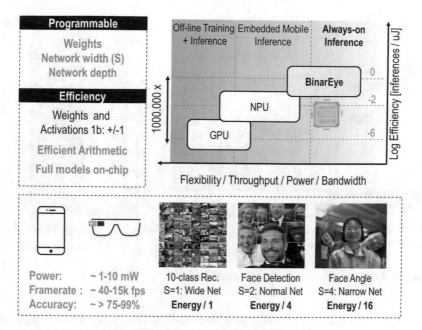

Fig. 6.2 Situating BinaryNet applications

that are used for network training, and large-scale cloud workloads. These are used for large networks consuming 100's of mJs per classification. The power budget in these systems is up to 100's of Watts. A second class is in mobile devices, such as smartphones and self-driving cars, with a power budget of Watts. In order to bring systems within this power budget, accuracy is typically kept high and only throughput is reduced compared to the HPC class. The envision processors in Chap. 5 are targetting this envelope. A final class performs always-on tasks within a 1-mW power budget. In order to achieve this, both throughput and accuracy should be lowered compared to the previous two classes, making BinaryNets a good fit for this class. The figure further illustrates the potential of BinaryNets, the target performance of the designed ASICs, and some examples of typical applications. The latter are further discussed in Sect. 6.5.

6.3   A Programmable Input-to-Label Accelerator Architecture

Figure 6.3 shows the top-level architecture of the visual wake-up ASIC architecture discussed in this chapter. The ASIC takes in raw 32×32 RGB image patches from an external source and outputs a classification label. The chip operates end-to-end or Input-to-Label (I2L), as all computations in this process are performed

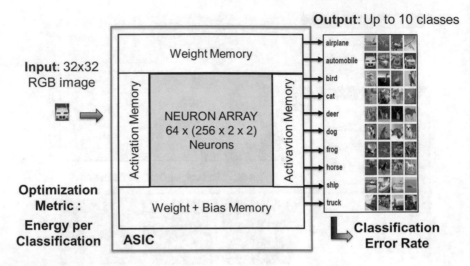

Fig. 6.3 System-level overview of the chip architecture

on-chip and there is no interaction with the outside world or external memory except for input and output data and some control signals. This I2L operation is a major difference with all currently published similar systems (Biswas and Chandrakasan 2018; Gonugondla et al. 2018; Khwa et al. 2018; Chen et al. 2018), which heavily rely on communication with off-chip memory or only implement parts of a full neural network. The figure also shows samples from CIFAR-10 (Krizhevsky and Hinton 2009), the benchmark used in designing and testing this chip.

In order to be able to offer this I2L operation, the chip has to support Binary CONV and MaxPool layers for feature extraction and FC layers for the final classification. Batch normalization is supported by transforming weights W and biases β to their equivalent forms W_{EQ} and β_{EQ} as in Sect. 6.1.

The main workload in the targeted BinaryNets is in the CONV layers; hence, the hardware implementing these will take most of the area of the chip. Figure 6.3 shows the baseline idea for these layers. The chip consists of an array of 64 reconfigurable binary neurons, surrounded by a total of 259 kB of weight SRAM on the north and south side storing full models and 2×32 kB of feature SRAM on the west and east side each storing full feature maps. Every neuron is a hardwired implementation of the triple summation in Eq. (6.2). Feature maps are processed on the array and ping-ponged from east-to-west and vice versa from layer-to-layer during classification. The north and south weight memories are used to update the weights and biases in the neuron array within a single layer, or from layer-to-layer. Not shown on this image is any control logic or the extra memory and hardware required to support FC layers. The rest of this section discusses the detailed operation of the chip. Section 6.3.1 discusses $256X$: a rigid baseline architecture used in the final Mixed-Signal chip implementation Sect. 6.4. Section 6.3.2 discusses SX: a more flexible extension to $256X$ based on the DVAFS ideas developed in Chap. 4.

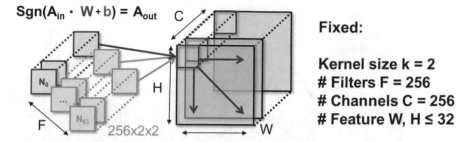

Fig. 6.4 Binary neuron operation and baseline network architecture

6.3.1 256X: A Baseline BinaryNet Architecture

Figure 6.4 illustrates the CONV operations that are performed in the $256X$ architecture, according to Eq. (6.1). Every layer is a series of $F \times k \times k \times C$ filters that are convolved over an input feature map of $W \times H \times C$. In the $256X$ architecture, F, C, and k are kept constant in order to simplify the logic and interconnect at the interface between memory and compute, hereby maximizing energy efficiency at the expense of flexibility. In every layer, F and C are fixed to 256, while the kernel size k is taken fixed as 2. This means that every neuron contains 1024 weights, equal to 128 bytes. Every layer hence requires 32 kB of weight storage. The size of the feature maps is limited to $W, H \leq 32$, which is sufficient for the targeted wake-up applications such as face detection. Throughout a network, the sizes of the intermediate feature maps reduce due to non-padded convolutions and max-pooling.

Figure 6.5 is a full overview of the $256X$ architecture. It is a hardwired array of 64 weight-programmable neurons surrounded by control units, drivers, decoders, and weight and activation memory. Also, on-chip is a separate unit that performs FC classification on binary generated features. The chip is connected to the outside world through an SPI interface that allows storing all configuration bits and network instructions. The weight buffers are written in a startup phase only through a separate scan interface. Input pixels are quantized into 85 levels per color ($6.4b\times$) and fed to the chip using a 16b-parallel input bus. On chip, they are then decoded to an 85-channel per RGB-channel thermometer format in order to fit into the 256-channel binary feature map format. This is further discussed below in Sect. 6.3.1.2. Finally, the output labels generated by the FC layer are transferred to the outside world on a 4b bus. This section discusses the operation of each of these blocks.

6.3.1.1 Neuron Array, Weight Updates, and Input and Output Demuxing

Figures 6.6 and 6.7 zoom in on more details of the binary neuron and on the mode of operation of the neuron array.

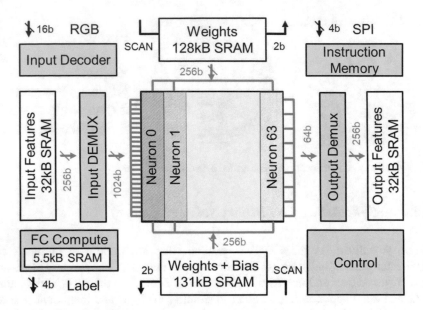

Fig. 6.5 Full 256X architecture overview

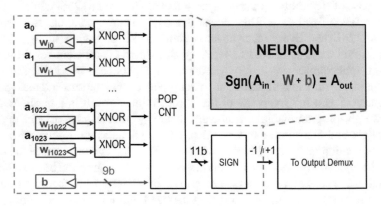

Fig. 6.6 A hardwired binary neuron

Figure 6.6 is the logical implementation of the core operation in Eq. (6.2). Each of the 64 neurons in Fig. 6.5 is hardwired to the illustrated structure. The neuron is a column of XNOR 1b-multipliers, followed by a POPCOUNT operator that combines all products with a 9b-filter bias. The full 11b sum is activated through a sign function before the filtering result is written back to feature memory. As one neuron is convolved over a full input image and weights can hence be reused, all weights are stored locally in a set of 1024 storage elements that can be either latches or flip-flops. Due to this local storage, neuron weights are reused up to $W \times H = 1024$ times in the input layer. As the POPCOUNT operator combines more than 1024 input bits, it will dominate the energy consumption in the neuron.

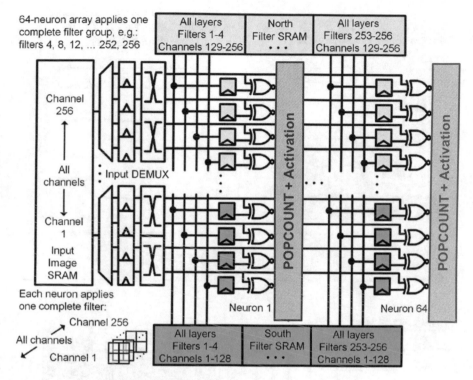

Fig. 6.7 256X neuron array architecture

Therefore, it is implemented efficiently in the analog domain in Sect. 6.4. Section 6.6 shows that the analog circuit is 4.2× more efficient than the conventional digital implementation.

Figure 6.7 shows how these neurons are physically implemented in the array and how they are connected to the surrounding weight and feature memory. The north weight memory supplies weights to the top halve of the neuron's weight flip-flops, while the south side supplies the bottom halve. Hence, the north side contains weights for channels 128–255, while the south side contains weights for channels 0–127 as well as all biases.

The neuron array contains only 64 neurons, while every layer exists out of 256 neurons. Hence, every layer is split into four groups and their convolution is time multiplexed. The neuron array always convolves one complete filter group, such as filter group 0 containing filter indices [0,4,8,...,252] or group 2 containing filter indices [2,6,10,...,254] of a specific CONVL. In a single CONVL, neuron weight updates or load (LD) sequences are interleaved with convolutional (CONV) sequences. After such an LD sequence, all weights of a neuron are kept local throughout a CONV cycle. This keeps the overall memory bandwidth to weight memory low. Hence, this architecture translates algorithm locality to reduced loading in the digital circuits. In order to process a layer with 256 neurons split into four groups, 4 LD-CONV cycles are needed on the 64-neuron array.

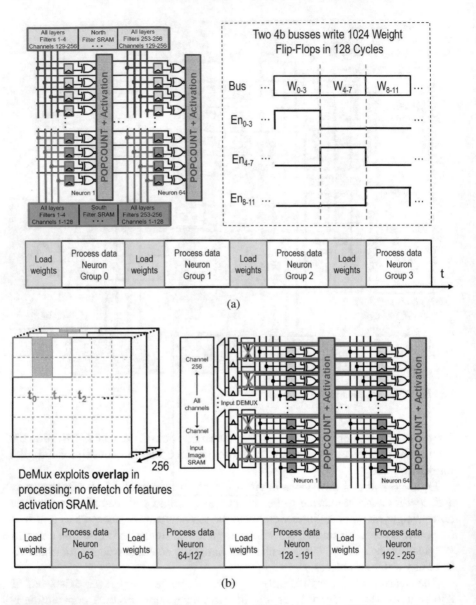

Fig. 6.8 Processing sequence in 256X. (**a**) Local WFF updates. (**b**) CONV processing with activation reuse

Figures 6.7 and 6.8a give more insight in how and when the weight flip-flops are updated. Per neuron, there are two 4-bit buses connected to all flip-flops, one bus for the north and one bus for the south side for a total 512b bandwidth per cycle. Every cycle, the 2×2 filter values for a single channel are written to the flops in both the north and south side of the neuron array. Hence, a single LD sequence,

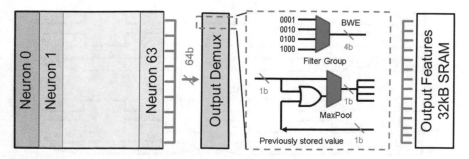

Fig. 6.9 Write back with optional max-pooling

writing the convolutional weights for 64 neurons to the array, takes 128 cycles plus the time to preload the biasing values that are stored on the bottom of the array. All the flip-flops are clock-gated on a per-row basis: every row has one integrated clock gating cell, allowing efficient activity reduction on these memory elements. The LD sequence overhead is only large when the chip is operating on small feature maps.

Figures 6.7 and 6.8b show how the neurons are connected to the feature map buffers through a demultiplexer (XMUX) that exploits data reuse in a Stride = 1 data flow. A single address in the feature map buffers stores the 256 channels associated with a feature map unit. When convolving the filter over a feature map, these values are fed sequentially to the array and reused across all 64 neurons, as they all operate on the same data. Figure 6.8b shows how XMUX buffers and reshuffles incoming data in order to limit the necessary SRAM bandwidth. For every Stride = 1 convolution step, only two additional units have to be fetched per channel. This is implemented by rewiring the connectivity of the input activations to the neurons, as illustrated in the figure. As the bandwidth to activation memory is limited to 256b, it takes one cycle to fetch one pixel and hence two cycles to fetch two pixels. Due to this scheme, the neuron array computes a result only once every two cycles, as it takes two cycles to ready a next set of input features. A further optimization would be to data-gate the neuron array such that its inputs only switch when the full $2 \times 2 \times 256$ image patch is valid. This would come at the cost of additional flip-flops and logic in the XMUX block.

Finally, as illustrated in Figs. 6.5 and 6.9, the activated output features are fed back to the feature buffer on the opposite side, again through an XMUX block. This block implements both standard and max-pooling operations. In the standard mode, 64-features are written to a 256-port SRAM when the correct result is available. A mask containing bit-write-enable (BWE) flags defined by the filter group number ensures that the 64 features are written to the correct channel positions, without overwriting other results.

The same block also implements streaming max-pooling as in Sect. 6.1 and shown in Fig. 6.1. Here, max-pooling is hence not implemented as an extra layer but is embedded at run-time into a CONVL. The only thing required is to modify the write-back scheme accordingly, as described below. In max-pooling

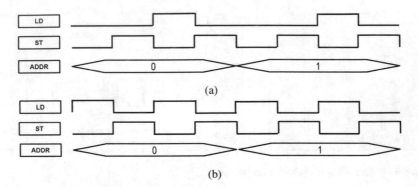

Fig. 6.10 Timing diagram for streaming max-pooling. (**a**) Even rows, and (**b**) odd rows. The address is computed as (column»1) + (row»1) × (width»1)

mode, depending on the r, c (row, column) indices of the current convolution, the previously written value for a given channel can be read-out first. If both r and c are even, the result is directly written to the correct address. If either r or c is uneven, before writing, the previously written result at a given address is read first, as shown in Fig. 6.10. The read/written address is computed as $addr = (c >> 1) + (r >> 1) \times (w >> 1)$, where w is the width of the input layer. The pre-fetched value is then compared (OR'ed) with the new incoming feature that belongs to the same max-pooling 2×2 cluster. Only if the incoming value is larger (equal to 1) than the previously stored value, the new feature will be written to the feature SRAM. Otherwise, the previous value is written again. The scheme described here is clearly not optimal. SRAM switching can be prevented by putting the BWE for the current channel, if the previously stored value is larger than or equal to the new feature. In that case, there is no need to rewrite the feature SRAM buffer.

6.3.1.2 Input Decoding

As discussed before, the neuron array is hardwired to operate on $W \times H \times 256$ 1b feature maps. This poses problems at the input layer, for two reasons. First, if input pixels are quantized to 1b, almost all information is lost in most cases. Second, raw input images are natively in an RGB $W \times H \times 3$ format, hence some adaptations have to be made to the input data to make it fit on the fixed-dimension neuron array, which requires feature maps to have 256 channels. Both issues are taken care of in a decoding block that transforms multi-bit input pixels into a format suitable for the array. More specifically, every channel R,G,B is represented as an 85b thermometer code, which is equivalent to a 6.4b-binary representation. Data transfer is always done in the binary format, to maintain a high IO speed. The final **feature vector** stored in memory is then a $3 \times 85 = 255 + 1b$ equivalent representation as in:

$$I_{256X} = [85'bI_{th}(R), 85'bI_{th}(G), 85'bI_{th}(B), 1'b1] \tag{6.9}$$

Here, the last bit is padding, but necessary in order for the vector to fit in the array. Through a right choice of equivalent weight kernel (see Eq. (6.11)), this padding bit does not influence the output of the network. All decoding discussed here is done on-chip at run-time.

The $2 \times 2 \times 3$ 1b **weight** kernel that is trained on the RGB input images should also be transformed into a format that fits in the neuron array. Note that in Eq. (6.1) the following replacement can be made through a simple rule of associativity:

$$I[c] \times W[f][c][i][j] = \left(\sum_{z=0}^{85} Iz[c] \right) \times W[f][c][i][j] =$$
$$\sum_{z=0}^{85} (Iz[c] \times W[f][c][i][j]) \tag{6.10}$$

where $\sum_{z=0}^{85} Iz[c]$ is the thermometer encoded representation of $I[c]$ this means that the 3-channel first layer filters can easily be expanded to a 256-channel format, by repeating each filter value 85 times, once for every element of the thermometer encoded input feature map. This is formalized below in Eq. (6.11):

$$W_{256X} = [85 \times W(R), 85 \times W(G), 85 \times W(B), mod2(i+j)] \tag{6.11}$$

Here, $85\times$ indicates that the same 1b weight is repeated 85 times. $mod2(i+j)$ makes sure that the redundant bit in the feature vector is canceled out, as i and j are the spatial indices as in Eq. (6.1).

6.3.1.3   Dense Layers

A sequence of a single FC or dense layer is supported in the $256X$ architecture, in order to facilitate input-to-label (I2L) behavior. Although this FC layer could potentially be implemented on the same neuron array as the one used for convolutional layers, here it is implemented separately outside of the array. This split allows separating the control and driving logic of the neuron array and optimizing it completely for the dominant CONVL's. As the FC layers are so small, the area overhead incurred by separation is limited. Apart from separate compute logic, it also uses a separate weight buffer. Figure 6.11 shows how this FC layer is implemented in a sequential way in a 64-way parallel processing unit. Every cycle, 64 weights and 64 features are transferred to the processing unit. Here, the binary values are multiplied and counted as in Eq. (6.4). A typical full feature map of $4 \times 4 \times 256 = 4096$ 1b features hence takes 64 cycles to complete a full vector

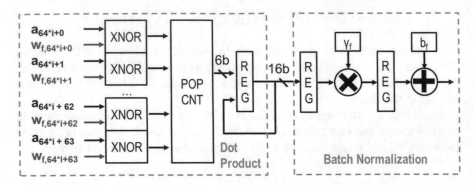

Fig. 6.11 Fully connected layer implementation: 64 activations and weights are fetched from memory every cycle

product for a single output. The partial sums are kept locally until the full sum is made. After, this result is multiplied with a filter-specific batch normalization factor γ_f after which an equivalent batch normalization bias $B_{EQ} = b_f$ is added. The trick to incorporate batch normalization completely into an equivalent weights set W_{EQ}, B_{EQ} as discussed in Sect. 6.1.2 is not possible here, as the sign activation function is not used. Finally, the layer can compare the output values for each class and can output the label of the largest value. The system does not support outputting a label and its confidence, which would be needed in a real detection application. Adding this functionality could be part of this project's future work.

The chip contains a total of 5.5 kB of FC-related SRAM (with weights, biases, and batch normalization parameters), in order to support a maximum of 10 output classes, to be classified based on $4 \times 4 \times 256$ feature maps.

6.3.1.4 System Control

$256X$ is built to be a programmable architecture, even though some of its functionality is hardwired. It is a non-Von Neumann processor with an instruction set, a program memory, and a complex data path as discussed above. An overview of the used instructions and the top-level control is given in Fig. 6.12. Figure 6.12a shows the three supported instruction types in $256X$ and some of their relevant parameters. All these instructions start a "layer" that can last for thousands of cycles. Only if the instruction is done, the program counter is increased. Three main instructions are designed:

- **IO** instructions are used to request or send encoded or raw input images. The instruction indicated the size of the data to be transferred, whether it requires decoding, and whether the data transfer is to/from the west or east activation memory.

Type	Options	Description
IO	W x H, raw or enc.	Input Layer, load and decode RGB images
CNN	Max-pooling, W x H	CONV, convolve 256 (256x2x2) filters on input
FC	W x H, up to 10 cat.	Generate output label through FC-layer

PM up to 16 instructions ≈ layers

(a)

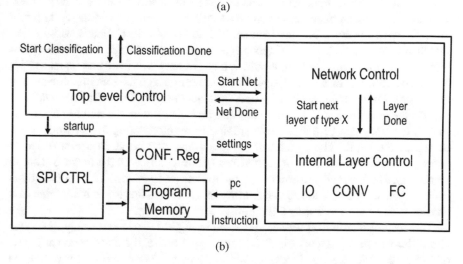

(b)

Fig. 6.12 Controlling the 256*X* architecture. (**a**) 256*X* instructions. (**b**) 256*X* control overview

- **CONV** instructions start and reconfigure a CONV layer. Relevant parameters are the size of the input feature map, whether or not max-pooling should be enabled, and whether features should be taken from the west or east activation memory.
- **FC** instructions start the final classification of a feature map into a classification label. The number of classes and the buffer containing the final activations (west or east) are variables in this instruction.

The system can be programmed through an SPI interface that connects to a configuration register file and the system's program memory. The latter is also implemented as a register file that stores up to 16 instructions or layers. In 256*X*, the total size of the weight memory limits the number of layers to 8 CONV and 1 FC layer.

Figure 6.12b gives a bird's eye overview of the system-level controller. A top-level controller controls the operating mode (startup, or running) and communicates with the outside world. The top-level controller masters all other elements on-chip but itself acts as a slave to another controller that is off-chip. In startup mode, both the weights can be updated through a scan interface and the SPI interface can be used to reconfigure the program memory. Whenever the system master asks for a classification, the top-level controller starts a new network. The network controller then loops through the instructions in the program memory. A new instruction is only started when the layer controller acknowledges that a layer is fully done.

6.3.2 SX: A Flexible DVAFS BinaryNet Architecture

SX is a more flexible architecture based on the rigid $256X$ baseline. The S in SX indicates that this architecture can support layers of different widths, as opposed to $256X$ where every layer has to have 256 filters of 256 channels wide. More specifically in SX, $F \times 2 \times 2 \times C$ filters can be convolved on $W \times H \times C$ feature maps. Here, F is still kept equal to C, but both values are not taken fixed and equal to 256, but can be $256/S$ with S = 1, 2, $or 4$. The need for a more flexible implementation is easily explained. If energy is to be traded for accuracy in $256X$, the only available knob is the network depth, which scales energy linearly with the number of layers. In SX, accuracy can be traded more efficiently by modulating the network width on top of the depth. The amount of computations, and hence in a first order its energy consumption, in a single layer scales quadratically with the factor S. Due to its rigidity, $256X$ would perform too many operations for some simple applications such as face detection. The added flexibility in SX solves this issue: it combines the efficiency of a hardwired array with more programmability.

This reconfigurability is further illustrated in Fig. 6.13, which visually illustrates two different operating modes in SX. In the S = 1 mode, the chip processes layers of 256 $2 \times 2 \times 256$ filters on $W \times H \times 256$ feature maps. As the neuron array contains only 64 neurons, its usage is time-multiplexed. Four LD-CONV cycles are needed to process a single layer on a single feature map. In the S = 4 mode, every layer uses 64 $2 \times 2 \times 64$ filters on $W \times H \times 64$ feature maps. As a single column in the array can now map four separate neurons, four feature maps can be computed in **parallel** using the same neuron array. On top of that, every layer only requires 64 neurons, so each layer is also 4× faster than in the S = 1 case, using only a single LD-CONV cycle per layer. The combination of these two effects leads to a speedup or equivalent energy reduction of 16×. On top of that, the 16× speedup can be exploited by using a lower frequency and supply voltage if throughput is kept constant, similar to the DVAFS technique discussed in Chap. 4. Although here, DVAFS is used in a sub-net parallel context rather than a subword-parallel context. More specifically, on CIFAR-10, a digital SX implementation achieves 86.05% accuracy in the S = 1 mode while consuming 15 uJ/classification. In the S = 4 mode, it achieves 76% accuracy while consuming 0.9 uJ/classification.

Figure 6.14 shows more details on how this can be implemented. It shows how the width of the CNN layers (F) and the number of channels of the input map (C) can be modulated with the variable degree of parallelism or batch size S as F = C = 256/S. Varying S hence allows trading energy in exchange for modeling accuracy. This is implemented by splitting neurons into four sub-neurons, each processing a $64 \times 2 \times 2$ dot product on 64 channels of the activation SRAM, as

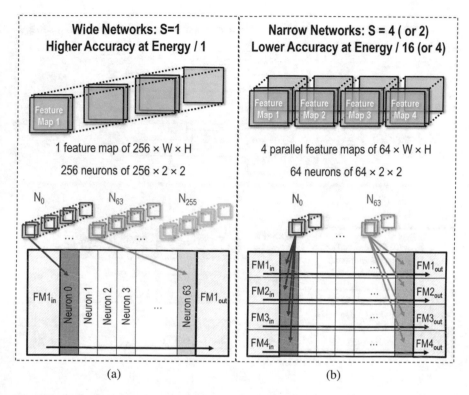

Fig. 6.13 Difference in operation in modes (**a**) $S = 1$ and (**b**) $S = 4$

shown in Fig. 6.14a. Depending on S, the intermediate outputs of the sub-neurons are combined into S output features. SX can hence either perform $F \times C \times k \times k$ = $256 \times 256 \times 2 \times 2$ (S = 1), $128 \times 128 \times 2 \times 2$ (S = 2) or $64 \times 64 \times 2 \times 2$ (S = 4) layers on one $256 \times W \times H$ (S = 1), two $128 \times W \times H$ (S = 2) or four $64 \times W \times H$ (S = 4) feature maps in parallel. This has a significant impact on throughput and energy, as further illustrated in Fig. 6.14b. If S = 1, a layer contains 256 neurons (F = C = 256), which means 4 LD-CONV cycles are required to process all neurons. If S = 4, a layer contains 64 (F = C = 64) neurons and requires only 1 LD-CONV cycle to process four input maps. Here, throughput and energy are improved quadratically with $S^2 = 16$, at the cost of reduced classification accuracy. This trade-off is further quantified in Sect. 6.5.

Figure 6.15 shows a top-level view of this reconfigurability. The SX architecture supports three different modes of "sub-net" parallelism. In the $N = 2$ mode, the chip is effectively split by partitioning the $256X$ architecture into two separate parts. This allows it to run smaller networks with a width of 128 filters on the same physical array. The $N = 4$ mode is the mode supporting the smallest networks with 64 filters per layer. In this case, the chip is reconfigured into four separate parts using the same physical array.

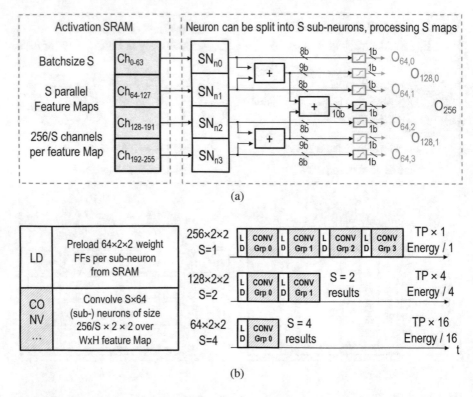

(a)

(b)

Fig. 6.14 SX requires a limited number of architectural changes compared to $256X$. (**a**) A neuron split into sub-neurons. (**b**) Timing in different modes

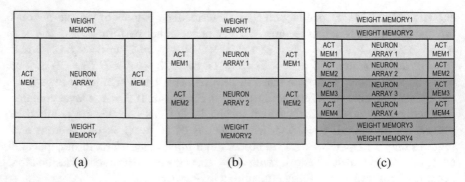

(a)   (b)   (c)

Fig. 6.15 System-level view of the three different operating modes supported in SX. (**a**) $S = 1$, id. $256X$. (**b**) $S = 2$. (**c**) $S = 4$

Apart from the discussed modifications above, moving from $256X$ to SX requires very little other changes. Obviously, the mode S should be stored in a configuration memory, which is a separate small register file in all implementations. Based on the value S in configuration memory, all the control loops and neurons are reconfigured in order to support the correct flow. However, the baseline functionality remains the same across modes S and only small changes are necessary. One major change is in the input decoding blocks. In $S = 1$, $255 = 85 \times 3$ channels are available per color. The possible input pixel precision drops in the $S = 4$ mode. Here, there are only $64 = 32 \times 3$ chip channels available to represent thermometer-coded RGB images. In general, the equivalent number of thermometer code channels per color is 85 in $S = 1$ (6.4b), 42 in $S = 2$ (5.5b), and 21 (4.5b) in $S = 1$. The number of possible input values per color is hence given by Eq. (6.12):

$$\#Values = \frac{256}{\#inputchannels \times S} \tag{6.12}$$

where $\#inputchannels$ equals three for RGB input images. This reduction in maximum input precision in the different modes will lead to a reduction in network performance in those modes. At this point, it is not clear how much of the loss in accuracy in modes 2 and 4 is due to this reduction in input precision, and how much is due to the smaller layers.

The design and measurement of a digital implementation of the SX architecture, a chip called BinarEye, is discussed in Sect. 6.5.

6.4 MSBNN: A Mixed-Signal 256X Implementation

With multiplication reduced to XNOR and memory access amortized across many computations by parallelism and reuse, wide vector summation in the neuron array becomes the main energy bottleneck. Because neural networks can tolerate a certain amount of random error in arithmetic computation without loss in classification accuracy, an opportunity exists to lower energy consumption by performing addition approximately with the internally analog, externally digital switched-capacitor (SC) neuron shown in Fig. 6.16 (Murmann et al. 2015; Moons et al. 2016). The SC neuron performs a weighted sum with data-dependent switching (apart from the comparator), hence its energy scales with activity factor like static CMOS logic. Precise matching between metal-oxide-metal (MOM) fringe capacitors in the 28-nm technology allows the unit capacitance to be as small as 1 fF without loss in classification accuracy. In contrast, the equivalent digital adder tree contains

---

The Bigger Picture

The SX BinaryNet architecture performs Dynamic-Voltage-Accuracy-Frequency-Scaling (Chap. 4) on the **network level**. Instead of operating in a subword-parallel mode at lower computational precision, this architecture operates in a sub-net parallel mode at lower benchmark accuracy. If the system-level throughput is kept constant, voltage and frequency can be modulated to increase energy efficiency.

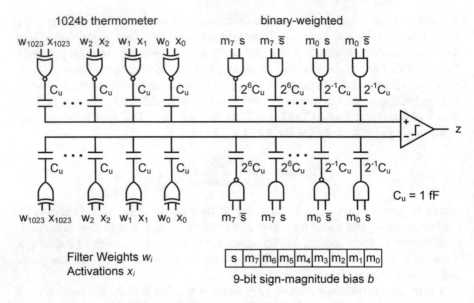

Fig. 6.16 Switched-capacitor neuron performs addition via charge redistribution

a large amount of switched capacitance across its multiple logic stages. The dominant noise source is the comparator, but its energy cost is amortized over 1024 weights. As a result, the SC neuron is amenable to low signal swing, and uses a combined 0.6 V analog reference and digital supply. The combination of activity factor proportionality, low voltage, and lower switched capacitance results in lower energy per classification using the SC neuron array relative to a digital neuron array implemented with static CMOS tree adders. This section discusses the mixed-signal binary CNN processor (MSBNN), an implementation of the 256X architecture achieving 3.8 µJ/86% CIFAR-10 I2L operation using the SC neuron array (Bankman et al. 2018).

6.4.1   Switched-Capacitor Neuron Array

The SC neuron in Fig. 6.16 performs a complete binary CNN filter computation according to:

$$z = 1; \tag{6.13}$$

XNOR gates perform point-wise products between weights w_i of a $2 \times 2 \times 256$ filter and activations x_i of a $2 \times 2 \times 256$ feature map patch. The addition operation is performed in the analog domain using a differential capacitive DAC (CDAC). The CDAC unit element is a 1-fF MOM fringe capacitor. A 1024b thermometer-coded

CDAC section adds point-wise products $w_i x_i$, and a 9b binary-weighted CDAC section adds the filter bias b, which is represented as a sign-magnitude integer. The 1b neuron output z is resolved using a voltage comparator. Because the SC neuron completes a binary CNN filter computation internally, the analog–digital interface is reduced to a single comparator decision.

Under ideal operation, the SC neuron computes z exactly according to (6.13). Three circuit non-idealities can potentially affect classification accuracy using the SC neuron array: unit-capacitor mismatch, comparator offset, and comparator noise. Detailed analysis shows that the actual, mismatched unit-capacitor values in the 1024b thermometer-coded CDAC section can be referred into a set of effective weights for the applied filter. Similarly, the unit-capacitor values in the 9b binary-weighted CDAC section and the comparator offset can together be referred into an effective bias. Unlike the random but static variations caused by mismatch, comparator noise is generated independently at every filter computation. Behavioral Monte Carlo simulations were run to determine the amount of comparator offset, noise, and unit-capacitor mismatch that the binary CNN can tolerate without degradation in classification accuracy, resulting in comparator offset and noise requirements of 4.6 mV and 460 μV, respectively, and a 1-fF unit capacitance.

6.4.2 Measurement Results

MSBNN was fabricated in TSMC's high-performance low-leakage (HPL) 28 nm CMOS technology, with the 1P8M_4X2Y1Z metal stack. Figure 6.19 shows a photo of the 2.44 mm×2.4 mm die. The design is split into four separate power domains: VDD for control and data transfer logic, VMEM for SRAM, VNEU = 0.6 V for the neuron array combined analog reference and digital supply, and VCOMP = 0.8 V for comparators. All measurements were taken at room temperature.

In the following measurements, MSBNN was tested on the full CIFAR-10 data set in order to build statistics characterizing noise and mismatch in the SC neuron array. The typical benchmark network used for these measurements is given in Fig. 6.20, which achieves 86.05% accuracy on the CIFAR-10 image classification task. The network contains 10 layers: 1 input layer, 8 CONV layers with max-pooling in layers 4 and 6, and 1 FC layer that outputs a 4b label. The full network requires 2 *billion* 1b operations, a task that is performed at an average I2L throughput of 23.7 fps/MHz or 47.6 GOPS/MHz. The chip is I2L, input-to-label, as it directly transforms a raw $32 \times 32 \times 32$ *RBG* image into one out of ten output category labels, without any interaction with other systems.

At nominal supply voltages in the digital periphery VDD = VMEM = 1.0 V, the chip operates at 380 FPS with $f_{CLK} = 16$ MHz and achieves 5.4 μJ/classification. Lowering VDD and VMEM to 0.8 V leads to 3.8 μJ/classification at 237 FPS with $f_{CLK} = 10$ MHz. Figure 6.17 shows energy per classification broken down by power domain at the operating point VDD = VMEM = 0.8 V. Not included in these energy

Fig. 6.17 Energy per classification of 3.8 μJ at the operating point VDD = VMEM = 0.8 V, 237 FPS. Breakdown by power domain

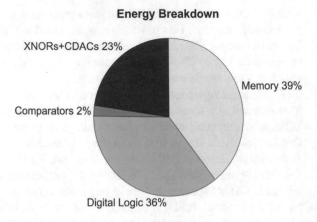

Fig. 6.18 Measured classification accuracy over 10 chips, 30 runs each through the 10,000 image CIFAR-10 test set

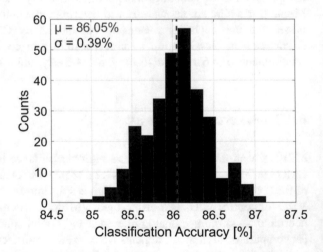

figures is the 1.8 V chip IO energy, which amounts to 0.43 μJ/classification (a small fraction of the core energy).

Figure 6.18 shows a histogram of classification accuracy measured across 10 chips, 30 runs each through the 10,000 image CIFAR-10 test set. The mean classification accuracy is 86.05%, the same as observed in a perfect digital model. The 95% confidence interval in mean classification accuracy is 86.01% to 86.10%. The standard deviation in classification accuracy is 0.39%, caused solely by noise and mismatch in the SC neuron (which can notably lead to a higher classification accuracy than in the perfect digital model) (Figs. 6.19 and 6.20).

Table 6.1 compares this work with prior art. On the same benchmark dataset (CIFAR-10), we achieve 40× improvement in energy per classification over the IBM TrueNorth chip, which does not exploit the locality of CNNs and hence suffers from high interconnect activity. The binarized DNN accelerator in Ando et al. (2017) has all memory on-chip, but because fully connected DNNs cannot exploit

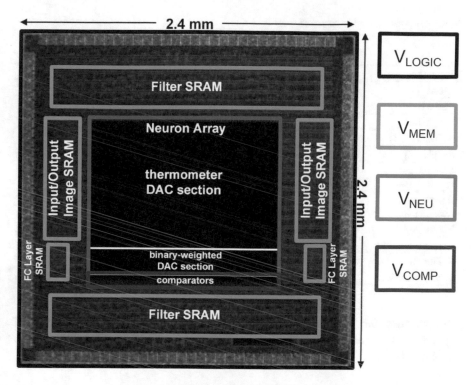

Fig. 6.19 Die photo of MSBNN, implemented in 28-nm CMOS. The chip is split into four voltage domains: one for all logic, all memory, one for the comparators, and one for the rest of the neurons. This granularity allows advanced DVFS methods to minimize energy consumption

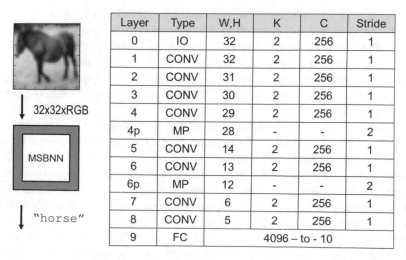

Layer	Type	W,H	K	C	Stride
0	IO	32	2	256	1
1	CONV	32	2	256	1
2	CONV	31	2	256	1
3	CONV	30	2	256	1
4	CONV	29	2	256	1
4p	MP	28	-	-	2
5	CONV	14	2	256	1
6	CONV	13	2	256	1
6p	MP	12	-	-	2
7	CONV	6	2	256	1
8	CONV	5	2	256	1
9	FC	4096 – to - 10			

Fig. 6.20 Typical 10-layer benchmark used to measure MSBNN. The network has 10 layers: 1 input layer, 8 CONV layers with max-pooling in layers 4 and 6, and 1 FC layer that outputs a 4b label. Using this network, the $256X$ architecture with periodic clearing achieves 86.05% accuracy on the test set of CIFAR-10 at 23.8 fps/MHz of I2L throughput

Table 6.1 Comparison to state of the art

	This work	IBM TrueNorth Esser et al. (2016)	VLSI '17 Ando et al. (2017)	VLSI '17 Buhler et al. (2017)
Technology	28 nm	28 nm	65 nm	40 nm
Algorithm	CNN	CNN	DNN	LCA
Dataset	CIFAR-10	CIFAR-10	MNIST	MNIST
Weight bits	1	1.6	1.6	4
Activation bits	1	1	1	1
Supply [V]	0.6, 0.8	1.0	0.55, −1.0	0.9
Classification accuracy [%]	86.05	83.41	90.1	88
Energy per classification [μJ]	3.79	164	0.28, −0.73	0.050
Power [mW]	0.899	204.4	50, −600	87
Frame rate [FPS]	237	1249	820 K, −3280 K	1.7 M
Arithmetic energy efficiency	532 1b-TOPS/W	–	6.0 − 2.3 TOPS/W	3.43 TOPS/W

weight reuse, the energy cost of an SRAM bit load is attached to each XNOR operation. The spiking LCA network in Buhler et al. (2017) exhibits low energy, but has relatively low accuracy for a lower-complexity task (MNIST).

6.4.3   Analog Signal Path Overhead

Analog circuit non-idealities must be managed in the SC neuron array in order to prevent loss in classification accuracy. This is accomplished using standard mixed-signal circuit techniques that do not contribute significantly to the SC neuron array energy per classification, which is dominated by the wires that connect its digital gates. We nonetheless describe the design overhead here in order to highlight the implementation differences between MSBNN and BinarEye, which was synthesized entirely in an RTL-to-GDSII standard cell flow.

The manner in which the SC neuron in Fig. 6.16 performs the weighted sum in (6.13) can be interpreted as a superposition of capacitive dividers. In order for the linear capacitive divider relation to hold true with respect to each CDAC bottom plate input, the charge at the CDAC top plate node must be zero. To accomplish this, convolution must be paused periodically while all CDAC bottom plate inputs are grounded and a switch connected to the CDAC top plate node (not shown in Fig. 6.16) discharges it. Such "clear" operations run every 100 to 300 clock cycles, as determined by the leakage current at the top plate node. At this rate, the contribution to energy per classification is negligible. Periodically pausing convolution for clear operations results in a 5.7% reduction in throughput relative to BinarEye, which operates at 25.1 fps/MHz.

The requirement on comparator offset can be significantly alleviated using calibration at startup. With 9 bits allocated to the filter bias, sufficient range exists to correct for comparator offset using the binary-weighted CDAC section, without an additional trim-DAC. The comparator offset is digitized at startup, stored in a local register, and subtracted from the bias loaded from SRAM during weight transfer. In environments where large temperature changes may induce significant offset drift, calibration can be performed periodically (e.g., once per second) at negligible cost in average energy per classification and throughput.

6.5   BinarEye: A Digital SX Implementation

Simultaneously with the MSBNN chip discussed in Sect. 6.4, an all-digital SX architecture was developed and measured. Ideally, MSBNN should have been SX as well. But, due to time constraints and tape-out deadlines, the team decided to stick with the vanilla $256X$ architecture for the mixed-signal design, as that design requires intensive manual layout, part of which was already done by the time SX was developed. However, there are no fundamental reasons why the mixed-signal version could not be implemented as SX. Both the mixed-signal neuron optimization and the added flexibility offered by SX can be regarded as completely orthogonal design efforts. The energy overheads in SX in the $S = 1$ mode where all filters are 256 wide are very low compared to the $256X$ baseline and can arguably be neglected. The only considerable overheads are in the multiplexing shown in Fig. 6.14a and in the more complex, multi-mode IO decompression. Hence, the performance of BinarEye for S $= 1$ is still comparable to the performance of MSBNN and can be used to estimate the gains of analog computation. The academic value of BinarEye is hence double: it is both a validation of the efficiency of the SX architecture and a case study in comparing analog-vs-digital computation.

Due to the added flexibility in SX, BinarEye has a wide energy-accuracy trade-off available by modulating not only the network depth but also the network width. For some tasks, such as face detection and face angle recognition, narrow networks are sufficient, allowing for much lower energy consumption. At low iso-accuracy on some benchmarks, the gains through the flexibility in SX outperform the gains through analog neurons in $256X$. This is further discussed in detail in Sect. 6.6.

6.5.1   An All-Digital Binary Neuron

The Neurons in BinarEye are implemented as in Fig. 6.14a, with a fully digital reconfigurable POPCOUNT operator. The POPCOUNT adder is synthesized from a tree description using the + operator in SystemVerilog using standard Synopsys Designware. Throughout the design, no further optimizations or custom interventions are done to minimize the energy consumption of the POPCOUNT operator.

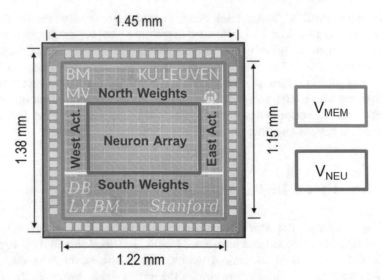

Fig. 6.21 Die photo of BinarEye. BinarEye is split into two separate power domains, in order to be able to separately scale supply voltages in the neuron array and the peripheral circuitry

The design is meant to be the baseline performance achievable by an experienced RTL designer, without any extra tools available.

6.5.2 Physical Implementation

Figure 6.21 is a die photo of the BinarEye chip, which was implemented in the same 28-nm CMOS technology as MSBNN but using the 1P8M-5X1Z1U metal stack. In this technology, BinarEye is implemented in less than $2\,mm^2$ as $1.45 \times 1.35\,mm^2$ with pads included. The system is synthesized and laid out to achieve 50 MHz operation at a nominal voltage of 1 V. All neurons and logic are automatically placed and routed with a minimum amount of constraints in order to give the optimization tools the maximum amount of freedom. As a top view, and ignoring the actual dimensions, this layout is nearly identical to the manual layout of MSBNN in Sect. 6.4. This chip, however, is only split into two power domains: one default domain for all peripheral logic, all control, and the SRAM memories and another one for the full neuron array. Figure 6.21 roughly indicates the placement of those power domains.

6.5.3 Measurement Results

This subsection discusses both the performance of the benchmark network of Fig. 6.20 and the system-level performance of multiple applications.

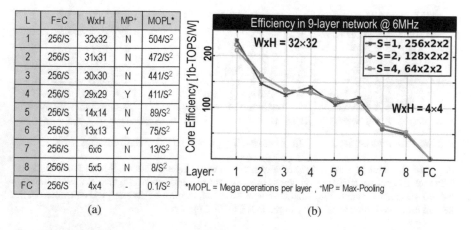

L	F=C	WxH	MP⁺	MOPL*
1	256/S	32x32	N	504/S²
2	256/S	31x31	N	472/S²
3	256/S	30x30	N	441/S²
4	256/S	29x29	Y	411/S²
5	256/S	14x14	N	89/S²
6	256/S	13x13	Y	75/S²
7	256/S	6x6	N	13/S²
8	256/S	5x5	N	8/S²
FC	256/S	4x4	-	0.1/S²

(a)

(b)

Fig. 6.22 Multi-mode benchmark network and performance in terms of 1b-TOPS/W for each of those modes. (**a**) Benchmark description in different modes. (**b**) Efficiency of all layers in the benchmark

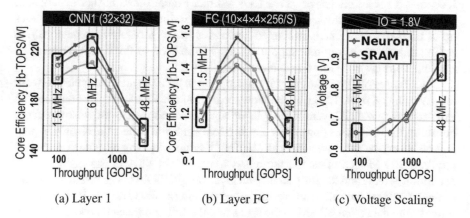

(a) Layer 1   (b) Layer FC   (c) Voltage Scaling

Fig. 6.23 Performance (1b-GOPS) of CONV layers (**a**) and of FC layers (**b**). Figure (**c**) shows how voltage can be scaled in the chip as function of frequency

6.5.3.1 Benchmark Network Performance

To demonstrate the chip's flexibility and the effectiveness of its different modes, Figs. 6.22 and 6.23 show measurement results for the layers of a typical 9-layer benchmark network for always-on imaging. Figure 6.22a shows the settings for all nine layers in the benchmarks in different modes. Depending on the scaling factor S, the width of each of the layers is reduced, also reducing the number of necessary operations considerably. When S = 1, the first layer in Figs. 6.22b and 6.23a receives a 32×32 7-bit RGB input and transforms it into a $256 \times 31 \times 31$ binary feature map in 500 M binary operations at core efficiencies up to 230 1b-TOPS/W at 6 MHz and 352 1b-GOPS. In the same layer, at S = 4, energy efficiency drops to

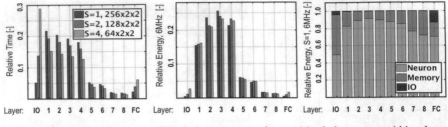

(a) relative time per layer   (b) relative energy per layer   (c) relative energy within a layer

Fig. 6.24 Relative (**a**) time and (**b**) energy spent on every layer within the 9-layer benchmark. (**c**) is the relative energy consumption across the different power domains. Figure (**c**) again motivates the move to analog neuron implementations in MSBNN, as most of the energy is consumed in the neuron array, and more specifically in the POPCOUNT operator

205 1b-TOPS/W at the same frequency. Figure 6.22b also shows that throughout the network, the core efficiency drops with smaller $W \times H$ feature maps, as the relative LD time increases compared to the CONV-time. Figure 6.23b further shows that, although BinarEye is not optimized for FC layers, it does achieve SotA efficiencies up to 1.5TOPS/W, which is similar to Ando et al. (2017). All measurements are at room temperature at optimal supply voltages from 0.66 to 0.9 V and frequencies ranging from 1.5 to 48 MHz, as indicated in Fig. 6.23c.

Some more details on the energy consumption in this 9-layer benchmark network are given in Fig. 6.24. These graphs show measurements for the relative computation time and energy of all layers of the 9-layer benchmark, as well as the relative energy consumption in all three power domains at 6 MHz in Emin, where the Neuron supply is 0.65 V and the memory supply 0.7 V. The "Memory" power domain contains all required SRAM memories: the north and south weight SRAM, the east and west activation SRAM and the fully connected layer weight SRAM. The "Neuron" power domain contains the full neuron array, all control logic, data-transfer logic used for transferring data from SRAM to the array, and the fully connected layer arithmetic. The "IO" power domain contains the IO ring at 1.8 V.

Figure 6.25 shows the simulated performance breakdown of BinarEye at $20\,MHz$ with the memory power domain at 1 V and the neuron array at 0.8 V. Almost 50% of the power consumption is due to wiring, while only 2% is in leakage at this frequency. This is achieved by primarily using HVT cells. As was stated before, the neuron array is dominant in power consumption, even at a non-nominal supply voltages. 75% of BinarEye's power consumption is in the neurons, of which more than 50% is in its wire connections. This motivates moving from an all-digital design to a mixed-signal design, as in Sect. 6.4.

6.5.3.2   Application-Level Performance

For several very simple applications, such as handwritten digit recognition and face detection, the 256-wide CONV layers in $256X$ and MSBNN are overkill and need

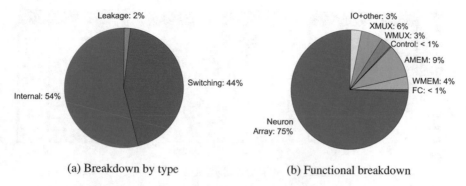

(a) Breakdown by type   (b) Functional breakdown

Fig. 6.25 Breakdown of simulated power consumption by type (**a**) and functional block (**b**)

too many operations and consume too much energy. Hence, those tasks can benefit from the support for narrow layers in BinarEye's SX architecture. Narrow networks require much less energy per classification but still have the classification capabilities for such simple tasks. This section discusses the performance of BinarEye on several applications in terms of I2L energy consumption and benchmark accuracy.

The system-level, IO-to-Label (I2L) performance and wide applicability of BinarEye are illustrated in Fig. 6.26. Here, the accuracy of several benchmarks is given as a function of S together with the energy consumption, throughput, and power for the 9-layer network of Fig. 6.22. All data sets and training scripts are available online at https://github.com/BertMoons/ (Moons et al. 2017c).

BinarEye can efficiently process several wake-up tasks relevant in the context of battery constrained wearable devices, among others. The numbers given in this paragraph and in Fig. 6.26 include the IO energy at 1.8 V. When S = 1 at 150 inferences/s and 14.4 uJ/inference, the chip achieves up to 86% on CIFAR-10, 98.1% precision / 95.7% recall on face and 98.2% precision / 83.3% recall on owner detection. If a slightly lower accuracy is allowed, the chip can scale up to 1700 inferences/s at its minimum energy point, consuming 0.92 uJ/inference at 94.5% precision on face detection. Once a face or owner is detected, the chip can reliably recognize up to 3–7 angles of the face, which can allow a mobile device to accurately track the relative position of a user's face relative to its own screen. If BinarEye is used as a detector in a sliding window fashion on QQVGA images, it achieves 1–20 fps at 1 mW and 15–200 fps at 10 mW using a stride of 16 pixels. At 1 mW, the chip hence offers up to 33 days of always operation on a typical 810-mWh AAA battery. How all this translates into face-detection benchmark accuracy is still being studied at this time.

6.5.4 DVAFS in BinarEye

The width-tuning knob allows for a wide range in network-level Dynamic-Voltage-Accuracy-Frequency-Scaling (DVAFS), a technique discussed in detail in Chap. 4.

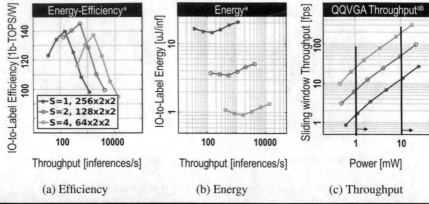

(a) Efficiency　　　　　　(b) Energy　　　　　　(c) Throughput

9-layer benchmark network performance								
Batch Size S	E/inf[c] [uJ]	T[c][infps / MHz]	P[c][uW/ MHz]	MNIST	CIFAR -10	Face Det. [d]	Owner Det. [e]	3 / 7 Angles
1	14.4	25	360	*98.85*	86	98.1*/95.7+	98.2*/83.3+	99.1/ 98.9
2	3.47	81	310	97.50	82	96.7*/94.2+	87.5*/89.4+	99.0 / 98.2
4	0.92	281	255	96.70	76	94.5*/87.6+	87.7*/87.1+	98.9 / 96.0

[a] Including 1.8V IO　　　　　　　　　　　　　　　[d] Tested on set of 5368 faces / 82665 backgrounds
[b] Sliding window with step size = 16, 9-layer network.　[e] Tested on set of 132 owner images / 1320 faces
[c] Including 1.8V IO, at minimum energy point.　　　* Precision, + Recall

(d) System-level Specs

Fig. 6.26 I2L system-level, including 1.8 V IO (**a**) efficiency, (**b**) energy per classification, (**c**) throughput, and (**d**) application-level specifications

In BinarEye, the network width is modulated, rather than the binary precision of the computational arithmetic. The energy per frame at constant throughput in this case can thus be described in a similar way as in classical DVAFS systems as in:

$$E \sim \frac{\alpha}{k_0 \times S} C \frac{f}{k_1 \times S} \left(\frac{V}{k_2}\right)^2 \tag{6.14}$$

Here, S is the level of sub-net parallelism as discussed in Sect. 6.5. Factors k_0 and k_1 are factors modeling the slight deviation of ideal throughput scaling with S, and k_2 is a scaling factor on the supply voltage the frequency is modulated. The impact of DVAFS in BinarEye can be seen from Fig. 6.27, which shows the I2L energy consumption for S = 1, 2, 4 at a constant throughput of 1000 inferences per second. A designer can hence choose to trade-off system-level energy for accuracy, depending on the requirements of the application. A network achieving 76% on CIFAR-10 consumes 20× less energy than a network achieving 86% accuracy. Figure 6.27 also compares the width-scaling knob in BinarEye, with the depth-scaling knob. At iso-accuracy, it is more energy efficient to modulate the network

Fig. 6.27 Energy and accuracy in multiple sub-net parallel DVAFS modes in BinarEye operating on CIFAR-10

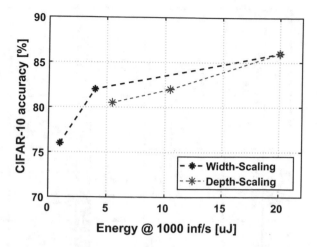

width than to modulate the depth. This is shown across the 80–86% accuracy range on CIFAR-10. The gains of depth-scaling compared to width scaling are significant, especially in the low-accuracy regions.

This kind of scalability is not necessarily useful within a single application. However, it can be used when a single computing platform should be able to cover applications of varying difficulty, such as face detection, owner recognition, and object recognition. In this case, all of those tasks can be run in the most efficient operating mode through DVAFS.

6.5.5   Comparison with the State-of-the-Art

Figure 6.28 compares BinarEye with other recent neural network implementations. YodaNN (Andri et al. 2016) is a binary-weight accelerator, as only the weights are 1b and activations are kept at high precision. Because of this, it achieves a very high accuracy on CIFAR-10, up to 91.7%. However, due to the large feature map sizes at 12b, the chip requires high bandwidth at run-time, which leads to an estimation of 1 mJ per classification if IO energy is included. Ando et al. (2017) is a binary/ternary DNN implementation that can only achieve 90% on MNIST, which is a very poor result, close to the accuracy of a single-layer linear classifier. As this chip fully optimized for DNN operation, it outperforms BinarEye up to a factor of 6 on FC layers, even in its 65-nm technology. In Moons et al. (2017b), accurate face detection is performed at 3 uJ/frame, only including the core energy. BinarEye achieves the same accuracy consuming only 0.9 uJ i2l, without any off-loading of intermediate data or weights. IBM's TrueNorth is about 10× less energy efficient than BinarEye, while using 500× more transistors to do the job. Bong et al. (2017) performs accurate face detection using an analog implementation of a Haar cascade. In 65 nm, it requires 12 uJ per classification, while BinarEye can achieve the same

	[2] TCAD '17	[3] VLSI '17	[Moo17] ISSCC'17	[Ess16] IBM TrueNorth	[Bon17] ISSCC'17	[Wha17] ISSCC'17	This work
Technology	65nm CMOS	65nm CMOS	28nm FDSOI	28nm CMOS	65nm CMOS	28nm CMOS	28nm CMOS
Frequency [MHz]	27 - 480	100 - 400	25 - 200	-	25	1200	1.5 - 48
Supply [V]	0.6 - 1.2	0.55 - 1	0.6 - 1	1	2.5	0.9	0.66 - 0.9
Active Area [mm²]	2.2	3.9	1.87	430	-	5.8	1.4
# of MACs	-	1728	256 - 1024	-	-	8	65536
Gate Count [NAND-2]	1.33M	-	1.95M	5.4B transistors	-	-	1.3M
On-Chip Memory	9.2kB SCM	51kB SRAM	148kB SRAM	51.2MB SRAM	80×20 An.	1158kB SRAM	328kB SRAM
#layers, #filters, sizes [-]	All, All, $<7\times7$	13, -, -	All, All, All	All, All, All	Haar-Filters	All	1-16, 64-256, 2×2
Supported Networks [-]	CNN	DNN	CNN	CNN	Analog	DNN	CNN + DNN
Precision [bits]	$1b \times 12b$	1-1.6b	1-16b	1-1.6b	1b	8b,16b	1b
Performance [GOPS]	15 - 377	345 - 1380	12 - 408	-	-	8.3	90 - 2800
Core* Eff. [TOPS/W]	58.6 - 9.6	6 - 2.3	10 - 0.3	-	-	0.58 - 0.95	230 - 145
I2L† Eff. [TOPS/W]	0.98 - 0.87	-	-	-	-	-	145 - 95

Acc. [%] / Emin [uJ/f]

Sub-column markers — [2]: [%], ↑†, ↑↑ | [3]: [%], ↑*, · | [Moo17]: [%], ↑†, ↑† | [Ess16]: [%], · | [Bon17]: [%], · | [Wha17]: [%], * | This work: [%], Core*, I2L†, S

	[2]	[3]	[Moo17]	[Ess16]	[Bon17]	[Wha17]	This work ([%] / Core* / I2L† / S)
MNIST	91.7 (↑†21, ↑↑1k)	90.1 (0.28)	-	83.4 (164)	-	97.5 (0.1)	97.4 / 0.2 / 0.21 / 4
CIFAR-10	-	-	-	-	-	60+ (3.8+)	86 / 13.82 / 14.4 / 1
Face Detection	-	-	94 (3)	-	>95 (11.8)	-	94.5 / 0.89 / 0.92 / 4
Owner Detection	-	-	-	-	-	-	98.2 / 13.82 / 14.4 / 1
7 Face Angles	-	-	-	-	-	-	98.2 / 3.4 / 3.47 / 2
Mode S	-	-	-	-	-	-	1ᵃ / 2ᵃ / 4ᵃ
Operations / Net [-]	1.2G	1.3M	12.4M	-	-	60k, 2.3M	2G / 0.5G / 0.12G
Inf/s @ Emin, Net [-]	15.6	205k	2.2k	1249	1	90k, 3.6	0.12k / 0.5k / 1.7k
EDP @Emin, Net [uJs]	64.1↑↑	1.4e-6*	1.4e-3↑*	0.131	-	1.1e-6, 0.001	1e-2† / 7e-3† / 5e-4†
P @ Emin, Net [mW]	15.760↑↑	50*	6.4↑*	204.4	-	22.4	2.2† / 1.8† / 1.6†

Fig. 6.28 Comparison of low-precision CNN processors on multiple benchmarks. Asterisk is the test core efficiency without IO power. Uparrow is the energy on Conv layers only. Alpha are numbers on this work is for a 5-layer network. Plus is extrapolated based on our own experiments. Dagger is with IO power included. Uparrow is the energy on Conv layers only. Alpha are numbers for the 9-layer benchmark of Fig. 6.22a. MNIST on this work is for a 5-layer network. Plus is extrapolated based on our own experiments

level of accuracy using less than 1 uJ. Finally, Whatmough et al. (2017) is $2\times$ more efficient than a narrow, 5-layer network ran on BinarEye at iso-accuracy on MNIST. However, the DNN's used by Whatmough et al. (2017) cannot be used to accurately model more complex CIFAR-10 images. At 0.92 uJ/classification, BinarEye still achieves more than 76% accuracy on CIFAR-10, while Whatmough et al. (2017) drops to 60%, even at similar or higher energy consumption.

6.6 Comparing Digital and Analog Binary Neural Network Implementations

This section explicitly compares the analog and digital versions of the $256X$ and SX architectures. Figure 6.29 is a full overview comparing some crucial specifications for both MSBNN and BinarEye chips, as well as extrapolated data for a fictitious mixed-signal and optimal digital SX implementation.

From measurements of the two chips, it is clear that the energy consumed in the MSBNN neuron is up to $12.9\times$ lower than in BinarEye, when the S $=1$ modes are directly compared on the same benchmark. This is due to the combined effect of the efficiency of the mixed-signal neuron, more advanced voltage scaling and a custom, very regular layout on an unconstrained area ($25\,\mathrm{k\mu m^2}$/neuron). In BinarEye on the other hand, the Neurons are a multi-stage tree structure that is irregularly laid out in a much more constrained area ($13\,\mathrm{k\mu m^2}$/neuron). Through simulations, the optimized digital neuron with a hand-designed adder tree laid out in the same area as MSBNN consumes $4.2\times$ more than the mixed-signal implementation, or $3.1\times$ less than the automated digital version. The difference in neuron efficiency is not only due to the design style (analog vs optimized digital vs digital) but also due to a difference in design constraints. The digital allows $2\times$ less area per neuron, which leads to an increase in wiring, specifically on the higher metals. On top of that, digital BinarEye was signed-off for 50 MHz, while the other implementations are signed-off for 20 MHz. This difference leads to increased buffering. Figure 6.29 also shows the DVAFS scalability of projected mixed-signal and optimized digital SX implementations. Here, we assume that the energy and throughput of the SC neuron array scale quadratically with S, as is the case for the digital neuron array. We additionally assume no degradation in mean classification accuracy for the 128X and 64X SC neuron array, as was the case in measurement for the 256X implementation. It should be noted that the assumption of quadratic scaling neglects the overhead of the filter bias CDAC section and the comparator, and that the assumption of no accuracy degradation has not been verified in simulation. With this, most of the previous conclusions hold here as well. However, as the discussed baseline versions are already more efficient, the gains of width scaling will be more modest than in a digital implementation. The maximum gain here is close to $10\times$, between the two ultimate modes, as compared to $15\times$ in BinarEye.

Chip name	BinarEye			MSBNN SOC	MSBNN SX SOC			OPTDIG SX SoC		
Design style	Automated Digital **measured**			Mixed-Signal **measured**	Mixed-Signal **estimated**			Optimized Digital **estimated**		
Technology	28nm CMOS (1P_8M, 5X1Z1U)			28nm CMOS (1P_8M, 4X2Y1Z)	28nm CMOS (1P_8M, 4X2Y1Z)			28nm CMOS (1P_8M, 4X2Y1Z)		
Active Area [mm²]	1.4			4.6	4.6			4.6		
Supported Modes	S=4	S=2	S=1	S=1	S=4	S=2	S=1	S=4	S=2	S=1
Supported Layers F × C	64×64	128×128	256×256	256×256	64×64	128×128	256×256	64×64	128×128	256×256
F @ I2L 478 GOPS [MHz]	13.5	10.6	9.5	10	14.2	11	10	14.2	11	10
Voltages @ F [V]	Neuron Array: 0.72 Memory: 0.7			Neuron Array: 0.6 Comp., Dig., Mem.: 0.8	Neuron Array: 0.6 Comp., Dig., Mem.: 0.8			Neuron Array: 0.6 Dig., Mem.: 0.8		
9-layer CIFAR-10 accuracy	76%	82%	86%	86.05%	76%	82%	86%	76%	82%	86%
9-layer FaceDet. Precision	94.5%	96.7%	98.1%	-	94.5%	96.7%	98.1%	94.5%	96.7%	98.1%
9-layer uJ/inference @ F	1.1	3.8	15	3.8	0.36	1	3.8	0.53	1.70	6.74
Neuron 9-layer uJ/inference @ F	0.84	3.1	12.1	0.94	0.1	0.3	0.94	0.27	1	3.89
I2L throughput @ F [inf/s]	3791	945	238	237	3791	945	237	3791	945	237
I2L core Power @ F [mW]	4.2	3.7	3.9	0.9	1.4	1	0.9	2.01	1.61	1.60
I2L core Eff. @ F [1b-TOPS/W]	122	132	134	532	366	520	532	237	297	299
I2L core Power @ 237 inf/s [mW]	0.35	0.9	3.9	0.9	0.2	0.3	0.9	0.21	0.48	1.60
6-layer MNIST accuracy	96.85%	97.4%	98%	-	96.85%	97.4%	98%	96.85%	97.4%	98%
6-layer uJ/inference @ F	0.65	2.17	8.2	-	0.2	0.54	2.1	0.31	0.95	3.73
Neuron 6-layer uJ/inference @ F	0.5	1.8	6.6	-	0.04	0.14	0.52	0.15	0.55	2.14
6-layer throughput @ F [inf/s]	5661	1642	455	-	5661	1642	455	5661	1642	455
I2L core Power @ F [mW]	3.27	3.17	3.7	-	1	0.8	0.95	1.76	1.56	1.70
I2L core Eff. @ F [1b-TOPS/W]	102	121	117	-	335	480	454	271	306	281
I2L core Power @ 455 inf/s [mW]	0.35	0.95	3.7	-	0.2	0.3	0.95	0.23	0.50	1.70

Fig. 6.29 Comparison of BinarEye, MSBNN, and the extrapolated performance of a fictitious mixed-signal and optimized digital *SX* implementation. **Estimated** numbers assume no accuracy degradation due to the analog non-idealities of the mixed-signal neuron array, as was the case in the measured mixed-signal chip

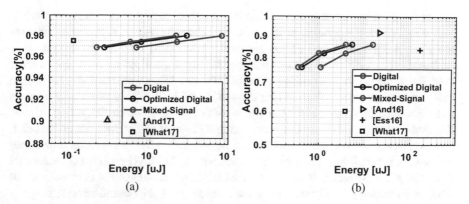

Fig. 6.30 Performance comparison of different versions of SX implementations vs the SotA. "Optimized digital" and "mixed-signal" are projections based on the measurements of MSBNN and BinarEye. (**a**) MNIST. (**b**) Performance comparison of different versions of SX implementations vs the SotA

Figure 6.29 not only compares the chips but also their performance in different operating modes. On MNIST, MSBNN can only lower its energy consumption by a factor of $<2\times$ by going to a shallower 6-layer network. In doing this, it also lowers its accuracy on the MNIST dataset. BinarEye has a separate knob, as it can also modulate the network width, which leads to a larger energy trade-off, without losing much more accuracy on simple data sets. In the S $= 4$ mode, it consumes only a third of the power of the MSBNN chip, at a slightly lower accuracy. This is the DVAFS effect, as in Chap. 4: for narrower networks, the chip operates in a sub-net parallel operating mode, allowing lower frequencies and supply voltages. At a constant throughput, in terms of classifications per second, it will then be more energy-efficient than modes with wider networks. For some simple applications, narrow networks can be sufficient to achieve high accuracy. This is most apparent when looking at the i2l core power at fixed throughput' rows in the table. On CIFAR-10, S $= 4$ leads to an accuracy loss of 10%. On MNIST, the accuracy drop is much lower at around 2%. The same observation holds for multiple other simple benchmarks in face detection and face angle recognition, as discussed in Sect. 6.5.3 and Fig. 6.26a. The performance of the different versions of BinarEye and the SX projections of an optimized digital and a mixed-signal implementation is further shown in Fig. 6.30a. The figure also shows the current SotA as a reference point. Figure 6.30b shows the performance on MNIST, while Fig. 6.30 shows the performance on CIFAR-10.

Even though it is less energy efficient, BinarEye has several advantages compared to MSBNN, specifically in an industrial environment. First, BinarEye works out-of-the-box and requires no offset or threshold calibration to compensate for comparator non-idealities, as is the case in MSBNN and was discussed in Sect. 6.4. BinarEye is a fully digital solution and can as such be operated in a mode where there are no deviations of the ideal functionality. In MSBNN, due to the analog nature of the computations, bits can flip under influence of thermal noise, as already happens at room temperature at low voltages. The influence of higher temperatures has not yet been measured on neither of the two chips. Finally, the design cycle of BinarEye is much shorter, much less challenging, and much more established than the design and verification of MSBNN. For BinarEye, only digital design tools and flows are required, while the design of MSBNN requires mixed-mode simulations and the manual design and layout of comparator circuitry, line drivers, flip-flops, latches, and so on. Once a working RTL-level netlist has been established, our experience shows that BinarEye can be fabricated by 1 experienced designer in about 2 months. Starting from the same netlist, MSBNN requires 5 months of development time before it can be fabricated. If the same system is to be built in another technology, a full redesign of the mixed-signal chip would be necessary.

6.7 Outlook and Future Work

Several innovations or optimizations are still to be made to BinaryNets in general and the $256X$ and SX architectures in specific to make them a good alternative to networks operating at higher precision.

Several doubts still exist on BinaryNets. These will have to be solved before BinaryNet hardware can be used in the wild.

- Although BinaryNets perform reasonably well on simple benchmarks such as CIFAR-10 and MNIST, it is not yet clear which real-world applications they can power. For example, even though BinarEye can achieve 86% accuracy on CIFAR-10 and general BinaryNets can be pushed up to 92%, they do not seem to outperform Viola–Jones based face detection in a vanilla face-detection pipeline. The IC-design research community should move away from treating benchmarks like CIFAR-10 or MNIST as representative of real-life applications and target full application pipelines instead.

Tips and Tricks

The analog implementation of MSBNN, which was laid out by hand, is close to $13\times$ more efficient than a tool generated digital layout in the same technology. The hand-designed digital adder is $3.5\times$ more efficient than the generated digital adder. In the tool-generated digital, every neuron is laid out on an average of $13\,\text{k}\mu\text{m}^2$, while the other versions have $2\times$ more ($25\,\text{k}\mu\text{m}^2$) area available. On top of that, the automated digital is signed-off for 50 MHz, while the others are signed-off for 20 MHz. The combination of these two effects—increased wire loads due to area constraints and increased buffering due to timing constraints—leads to the large difference in efficiency. This is a lesson indicating the importance of both constraints and their effect on energy consumption.

- Very little research efforts have gone into testing binary or quantized neural networks on more advanced network architectures such as ResNets, DenseNets, and MobileNets. Preliminary experiments on ResNet show discouraging results, as it does not support BinaryNets, for example (see Chap. 3). This is a problem, as most of the more novel network architectures not only achieve higher accuracies but do this using less weights and less operations. This raises the question if advanced quantization might be a wrong answer to the right question: how to minimize energy consumption in the state-of-the-art neural networks.

For the designed $256X$ and SX architectures, several changes can be made to make them more energy-efficient.

- As discussed in Sect. 6.3, the bandwidth from the west and east feature buffers is too low to support a CONV operation every cycle. In the current version, the filters are shifted and computed only once every two cycles.
- To maximize the impact of the previous optimization, the bandwidth from the memory-like neuron array to the north and south weight buffers could be increased as well in order to speed up the weight-LD cycles. A higher bandwidth would reduce the time overhead of the LD cycle and increase overall throughput.
- The current $256X$ and SX architectures do not support input images larger than $32 \times 32 \times RGB$. The architectures could be expanded for usage on larger inputs. This can be done in two complementary ways. First, the feature map buffers can easily be quadrupled or octupled in size, without major impact on energy efficiency. On top of that, the chip could be embedded into a larger system with external DRAM, to support arbitrarily large feature maps.
- The current array is hardwired to $256/N \times 2 \times 2$ neurons. It could pay off to generalize the architecture to 1×1 and depthwise separable 3×3 layers. Going to $256/N \times 3 \times 3$ filters would only require minor architectural changes. However, in our experiments, this does not lead to better results on CIFAR-10, but might on other benchmarks or on larger feature maps.
- The same SX architecture can be expanded to support 2b or 4b (or Nb for that matter) operators. As precision is increased, the gap in energy efficiency between mixed-signal and digital processing closes (Murmann et al. 2015). According to Chap. 3, specifically 4b networks are generally a better trade-off in energy vs accuracy. On top of that, they are proven to be compatible with ResNets as well in Moons et al. (2017a).
- In terms of layout, several improvements can be made as well, both in digital BinarEye and in the MSBNN. In the current versions, feature memories are placed left and right of the array, which makes the lines going from left to right longer than they have to. Placing the memories and all horizontal line drivers in the middle of the array would shorten those lines and minimize the now significant wiring energy. This strategy would be similar to the one used in classical SRAM macros.

6.8   Conclusion

Due to their complexity, CNNs have not yet been deployed on always-on mobile platforms such as smartphones, where an always-on gesture, face, owner, and angle detector can be used as a wake-up sensor for its screen or application processor.

This chapter discusses the architecture and hardware design, measurements, and application-level performance of two BinaryNet acceleration ASICS that could achieve just this. They target a wide variety of always-on visual wake-up applications on battery-constrained mobile devices, typically the earliest stages in the hierarchical processing cascades discussed in Chap. 2. This goal of always-on compute is achieved through: (1) extremely efficient hardware in the analog domain, (2) maximal reuse of weights in a memory-like neuron array (3), storing full models and feature maps on chip, requiring no off-chip bandwidth, and (4) flexibility on three levels: retrainable weights, a programmable network depth, and a reconfigurable network width. As part of a larger team at Stanford, we developed two hardware architectures for these BinaryNets that can operate in an always-on fashion. $256X$ is implemented in a mixed-signal fashion in MSBNN. SX is implemented in the all-digital BinarEye chip.

MSBNN is a mixed-signal binary CNN implementation of $256X$, that performs image classification of moderate complexity (86% on CIFAR-10) in the analog domain (1) and employs near-memory data processing (2,3) to achieve peak 532 1b-TOPS/W at a classification energy of 3.8 uJ, a 40× improvement over TrueNorth (Esser et al. 2016).

BinarEye is an all-digital SX implementation that focuses on improved flexibility. Its computations are fully digital, but its architecture expands on $256X$ by allowing more reconfigurability in its network mapping (4). Because of this, BinarEye can map a wide range of applications, while offering an input-to-label, full-system efficiency up to 145 1b-TOPS/W. This allows running 1k inferences/s of 125M operations each at 1 mW or 1 uJ/inference, while still achieving 94% precision in face detection and 90% accuracy in multiple other tasks. BinarEye outperforms (Ando et al. 2017; Moons et al. 2017b; Andri et al. 2016; Bong et al. 2017) up to 70× on CIFAR-10 at a slightly lower accuracy and 10× the throughput and by 3.3–12× on face detection at iso-accuracy.

Due to its computations in the analog domain, MSBNN consumes almost 13× less energy in its neurons than BinarEye. A projection of a hand-designed digital neuron implementation consumes 3.1× less neuron energy than BinarEye. On the system level, MSBNN is 4× and 1.76× better than BinarEye and the hand-designed digital projection, respectively.

References

Ando K, Ueyoshi K, Orimo K, Yonekawa H, Sato S, Nakahara H, Ikebe M, Asai T, Takamaeda-Yamazaki S, Kuroda T, et al (2017) Brein memory: a 13-layer 4.2 k neuron/0.8 m synapse binary/ternary reconfigurable in-memory deep neural network accelerator in 65 nm CMOS. In: Symposium on VLSI circuits, 2017. IEEE, pp C24–C25

Andri R, Cavigelli L, Rossi D, Benini L (2016) Yodann: an ultra-low power convolutional neural network accelerator based on binary weights. In: IEEE computer society annual symposium on VLSI (ISVLSI), 2016. IEEE, pp 236–241

Bankman D, Yang L, Moons B, Verhelst M, Murmann B (2018) An always-on 3.8 uj/classification 86accelerator with all memory on chip in 28nm CMOS. In: International Solid-State Circuits Conference (ISSCC) technical digest

Biswas A, Chandrakasan A (2018) Conv-ram: an energy-efficient SRAM with embedded convolution computation for low-power CNN-based machine learning applications. In: International Solid-State Circuits Conference (ISSCC)

Bong K, Choi S, Kim C, Kang S, Kim Y, Yoo HJ (2017) 14.6 a 0.62 mw ultra-low-power convolutional-neural-network face-recognition processor and a cis integrated with always-on haar-like face detector. In: IEEE International Solid-State Circuits Conference (ISSCC), 2017. IEEE, pp 248–249

Buhler F, Brown P, Li J, Chen T, Zhang Z, Flynn M (2017) A 3.43 tops/w 48.9 pj/pixel 50.1 nj/classification 512 analog neuron sparse coding neural network with on-chip learning and classification in 40 nm CMOS. In: Symposium on VLSI circuits, pp 30–31

Chen WH, Li KX, Lin WY, Hsu KH, et al (2018) A 65 nm 1 mb nonvolatile computing-in-memory reram macro with sub-16ns multiply-and-accumulate for binary DNN ai edge processors. In: IEEE International Solid-State Circuits Conference (ISSCC), IEEE

Courbariaux M, Bengio Y, David JP (2015) Binaryconnect: training deep neural networks with binary weights during propagations. In: Cortes C, Lawrence ND, Lee DD, Sugiyama M, Garnett R (eds) Advances in neural information processing systems, vol 28. Curran Associates, Inc., Red Hook, pp 3123–3131

Esser S, Merolla P, Arthur J, Cassidy A, et al (2016) Convolutional networks for fast, energy-efficient neuromorphic computing. In: Proceedings of the national academy of sciences

Gonugondla SK, Kang M, Shanbhag N (2018) A 42pj/decision 3.12tops/w robust in-memory machine learning classifier with on-chip training. In: International Solid-State Circuits Conference (ISSCC)

Han S, Mao H, Dally WJ (2016) Deep compression: compressing deep neural network with pruning, trained quantization and huffman coding. In: International Conference on Learning Representations (ICLR)

Hubara I, Courbariaux M, Soudry D, El-Yaniv R, Bengio Y (2016) Binarized neural networks. In: Advances in Neural Information Processing Systems (NIPS)

Ioffe S, Szegedy C (2015) Batch normalization: accelerating deep network training by reducing internal covariate shift. arXiv preprint:150203167

Khwa WS, Chen JJ, Li JF, Si X, et al (2018) A 65 nm 4 kb algorithm-dependent computing-in-memory SRAM unit-macro with 2.3 ns and 55.8 tops/w fully parallel product-sum operation for binary DNN edge processors. In: IEEE International Solid-State Circuits Conference (ISSCC). IEEE

Krizhevsky A, Hinton G (2009) Learning multiple layers of features from tiny images. Technical report

LeCun Y, Cortes C (1998) The MNIST database of handwritten digits

Moons B, De Brabandere B, Van Gool L, Verhelst M (2016) Energy-efficient convnets through approximate computing. In: Proceedings of the IEEE Winter Conference on Applications of Computer Vision (WACV), pp 1–8

Moons B, Goetschalckx K, Van Berckelaer N, Verhelst M (2017a) Minimum energy quantized neural networks. In: Asilomar conference on signals, systems and computers

Moons B, Uytterhoeven R, Dehaene W, Verhelst M (2017b) Envision: a 0.26-to-10 tops/w subword-parallel dynamic-voltage-accuracy-frequency-scalable convolutional neural network processor in 28 nm FDSOI. In: International Solid-State Circuits Conference (ISSCC)

Moons B, et al (2017c) Bertmoons github page. http://github.com/BertMoons, Accessed: 01 Jan 2018

Moons B, Bankman D, Yang L, Murmann B, Verhelst M (2018) Binareye: an always-on energy-accuracy-scalable binary CNN processor with all memory on-chip in 28 nm CMOS. In: IEEE Custom Integrated Circuits Conference (CICC)

Murmann B, Bankman D, Chai E, Miyashita D, Yang L (2015) Mixed-signal circuits for embedded machine-learning applications. In: 2015 49th Asilomar conference on signals, systems and computers, pp 1341–1345. https://doi.org/10.1109/ACSSC.2015.7421361

Rastegari M, Ordonez V, Redmon J, Farhadi A (2016) XNOR-net: Imagenet classification using binary convolutional neural networks. In: European conference on computer vision. Springer, Berlin, pp 525–542

Russakovsky O, Deng J, Su H, Krause J, Satheesh S, Ma S, Huang Z, Karpathy A, Khosla A, Bernstein M, et al (2015) Imagenet large scale visual recognition challenge. Int J Comput Vis 115(3):211–252

Sze V, Yang TJ, Chen YH (2017) Designing energy-efficient convolutional neural networks using energy-aware pruning. In: Conference on Computer Vision and Pattern Recognition (CVPR)

Whatmough PN, Lee SK, Lee H, Rama S, Brooks D, Wei GY (2017) 14.3 a 28 nm soc with a 1.2 ghz 568 nj/prediction sparse deep-neural-network engine with> 0.1 timing error rate tolerance for IOT applications. In: IEEE International Solid-State Circuits Conference (ISSCC), 2017. IEEE, pp 242–243

Zhao R, Song W, Zhang W, Xing T, Lin JH, Srivastava M, Gupta R, Zhang Z (2017) Accelerating binarized convolutional neural networks with software-programmable FPGAS. In: Proceedings of the 2017 ACM/SIGDA international symposium on field-programmable gate arrays. ACM, New York, pp 15–24

Zhu C, Han S, Mao H, Dally WJ (2016) Trained ternary quantization. arXiv preprint:161201064

Chapter 7
Conclusions, Contributions, and Future Work

This dissertation has focused on techniques to minimize the energy consumption of deep learning algorithms for embedded applications on battery-constrained wearable edge devices. Although SotA in many typical machine-learning tasks, deep learning algorithms are very costly in terms of energy consumption, due to their large amount of required computations and huge model sizes. Because of this, deep learning applications on battery-constrained wearables have only been possible through wireless connections with a resourceful cloud. This setup has several drawbacks. First, there are privacy concerns. This setup requires users to share their raw data—images, video, locations, speech—with a remote system. As most users are not willing to share all of this, large-scale applications cannot yet be developed. Second, the cloud-setup requires users to be connected all the time, which is unfeasible given current cellular coverage. Furthermore, real-time applications require low latency connections, which cannot be guaranteed using the current communication infrastructure. Finally, this wireless connection is very inefficient—requiring too much energy per transferred bit for real-time data transfer on energy constrained platforms. All these issues—privacy, latency/connectivity and costly wireless connections—can be resolved by moving towards computing on the edge.

Computing in the edge is only possible if these deep learning algorithms can be run in a more energy-efficient way, within the energy- and power-budget of the computing platform available on a wearable device. In order to achieve this, several innovations are necessary on all levels of an application's design hierarchy. Smarter **applications** can be developed for more statistically efficient deep learning **algorithms**, which in turn should run on optimized **hardware** platforms built on specifically tailored **circuits**. Finally, designers should not focus on any of these fields separately, but should **co-optimize** hardware and software to create minimum energy deep learning systems. I've tried contributing to all of these factors in this dissertation.

© Springer Nature Switzerland AG 2019
B. Moons et al., *Embedded Deep Learning*,
https://doi.org/10.1007/978-3-319-99223-5_7

The conclusions of all previous chapters are now briefly reconsidered, followed by an overview of suggestions for future work.

7.1 Conclusions

Chapter 2 focuses on the **application level**. The chapter generalizes wake-up systems to multi-stage hierarchical cascaded systems that have a lower system-level cost and are better adapted to skewed data with non-uniform probability distributions. A design framework is presented that can be used to simultaneously optimize performance and minimize costs. Linked to this is a theoretical roofline model providing insight in the performance of the individual stages in the hierarchy. General trends in hierarchical cascades are derived through analyzing a generic example. It is shown that, while hierarchical cascades do not bring significant benefits for uniform input data statistics, systems with skewed input data statistics, such as speech and object detection tasks, benefit from deeper cascades. The approach is further illustrated by designing a four-stage 100 face recognition application. An optimal operating point exists, where four orders of magnitude in cost-efficiency can be gained compared to the single-stage classifier and two orders of magnitude compared to the traditional two-stage wake-up-based system.

Chapter 3 is both an introduction to various SotA means of hardware–software co-optimizations in the context of deep learning and a deeper dive in one specific method that was developed in this book: energy-efficient networks through exploiting their computational and hardware-level fault-tolerance. This is done by reducing the internal computational precision of the networks. The distinction is made between two approaches: test-time fixed-point neural networks (FPNNs) that are the result of an advanced fixed-point analysis on a floating-point pre-trained network and train-time quantized neural networks (QNN) that are trained from scratch in fixed-point. Because QNNs are explicitly trained at low precision, they outperform FPNNs in terms of energy-efficiency. It is shown optimal 4b quantized neural networks are generally optimal on the tested benchmarks, 3–10× more efficient than 8b and 16b implementations. Moving to nonlinearly quantized neural networks, using the approach first discussed in Han et al. (2016), is up to 10× more efficient than linear quantization at high accuracy.

Chapter 4 focuses on dynamic-voltage-accuracy-frequency (DVAFS), an effective dynamic circuit-level technique in approximate computing. This technique allows converting the fault-tolerance of deep learning algorithms into maximum energy savings. To the best of our knowledge, DVAFS achieves higher block- and system-level energy gains than any other approximate computing technique at reduced precision. It outperforms the static and dynamic approximate computing state of the art up to 3.5× on the block level and up to 4.5× in a parallel SIMD-processor. DVAFS offers the opportunity to lower power and energy consumption per execution for the whole system as it modulates frequency for the full system at low precision while keeping throughput constant. This allows reducing all non-compute overheads in a real approximate computing system. None of the previous works have taken this into account, making DVAFS a major contribution to the field. Finally, both the functional (subword-parallel building blocks) and physical (enforcing critical path scaling) implementation challenges of DVAFS are covered.

Chapter 5 discusses Envision: two generations of energy-accuracy scalable, sparse convolutional neural network processors, based on the techniques discussed in Chaps. 3 and 4. Its scalability makes it suitable for usage in a hierarchical scenario, as discussed in Chap. 2. The focus of Envision is on mid-range power consumption, within the 10–300 mW mobile power range. These processors offer (A) an energy-efficient baseline in a parallel 2D-MAC-array architecture. They can minimize energy consumption depending on (B) the sparsity of a CNN-application and (C) its computational precision requirements through DVA(F)S. The combination of these three techniques led to energy-efficiencies up to 2.6 and 10 TOPS/W at 76 GOPS for sparse / low-precision workloads, which made Envision the most efficient CNN implementations in 2016 (Moons and Verhelst 2016) and 2017 (Moons et al. 2017). DVA(F)S compatibility leads to $20\times$ $(8\times)$ efficiency gains at low precision compared to the 16b full precision baseline. Depending on the application, this does not lead to a loss in system-level accuracy. A spatial sparsity-skipping scheme leads to 2–$2.7\times$ reductions in energy, if input datastreams are 50–90% sparse.

Chapter 6 discusses the architecture and hardware design, measurements, and application-level performance of two BinaryNet acceleration ASICs. These chips target a wide variety of always-on visual wake-up applications on battery-constrained mobile devices within a 1–10 mW power budget, typically the earliest stages in the hierarchical processing cascades discussed in Chap. 2. This goal of always-on compute is achieved through: (1) extremely efficient hardware in the analog domain, (2) maximal reuse of weights in a memory-like neuron array (3), storing full models and feature maps on chip, requiring no off-chip bandwidth, and (4) flexibility on three levels: retrainable weights, a programmable network depth, and a reconfigurable network width. As part of a larger team at Stanford, we developed two such hardware architectures. $256X$ is implemented in a mixed-signal fashion in MSBNN. SX is implemented in the all digital BinarEye chip.

MSBNN is a mixed-signal binary CNN implementation of $256X$ that performs image classification of moderate complexity (86% on CIFAR-10) in the analog domain (1) and employs near-memory data processing (2,3) to achieve peak 750 1b-TOPS/W at a classification energy of 3.8 uJ, a $40\times$ improvement over TrueNorth (Esser et al. 2016).

BinarEye is an all digital SX implementation that focuses on improved flexibility. Its computations are fully digital, but its architecture expands on $256X$ by allowing more reconfigurability in its network-mapping (4). Because of this, BinarEye can map a wide range of applications, while offering an Input-to-Label, full-system efficiency up to 145 1b-TOPS/W. This allows running 1k inferences/s of 125M operations each at 1 mW or 1 uJ/inference, while still achieving >94% precision in face detection and >90% accuracy in multiple other tasks. BinarEye outperforms (Ando et al. 2017; Moons et al. 2017; Andri et al. 2016; Bong et al. 2017) up to 70× on CIFAR-10 at a slightly lower accuracy and 10× the throughput and by 3.3–12× on face detection at iso-accuracy.

The work in this chapter is hence also a comparison between analog and digital compute. Due to its computations in the analog domain, MSBNN consumes almost 13× less energy in its neurons than BinarEye. A projection of a hand-designed digital neuron implementation consumes 3.5× less neuron energy than BinarEye. On the system level, MSBNN is 4× and 1.7× better than BinarEye and the hand-designed digital projection, respectively.

This work also allows to draw more general conclusions:

- In order to bridge the efficiency-gap between deep learning algorithms and mobile and always-on power budgets, applications, algorithms, and hardware platforms will have to be co-optimized. Optimizations on the application, algorithm, and circuit level show a 100× (cascading), 10× (optimal QNNs), and 10–15× (sparse DVAFS in an optimized architecture) or 3–40× (BinaryNets) improvement over naive implementations.
- Any optimized hardware platform for deep neural networks should exploit (A) an optimized parallel architecture, (B) the high degrees of sparsity in any neural network and (C) exploit its fault-tolerance on the circuit level. The most efficient technique to do the latter is dynamic precision scaling.
- Co-optimizing deep learning algorithms to specific constrained hardware platforms can lead to large efficiency gains. By investing in low-precision algorithmic techniques, a whole range of circuit-level techniques become feasible. The BinaryNet implementation in Chap. 6 is just an example, but there are a whole range of other hardware techniques—optical computing, computing-in-memory, computing-in-flash, and RRAM—that can offer orders of magnitude increase in compute efficiency if deep learning algorithms were to be tailored to their needs.
- Dynamic bit width scalability, as in DVAFS, leads to high energy-scalability in processor architectures, but also incurs a lot of overhead. It would pay off to limit the supported range of bit widths, maintaining scalability at a much lower cost.
- Binary neural networks are interesting case studies, but unproven on large-scale networks and benchmarks and even real applications. Other implementations that use binary or quantized CNN kernels combined with specific affine output transformations (such as XNOR-net) seem more promising as they also achieve high accuracy on more complex data sets.
- Small neural network benchmarks, such as the ones primarily used in this book, are not representative for the real world. The IC-design community should move from proving their prototypes towards either larger data sets, such as IMAGENET, or to real-world applications.

7.2 Suggestions for Future Work

- The work of Chap. 2 can move in the direction of the work proposed in Huang et al. (2017). Here, early stopping classifiers are added after the first couple of layers in a CNN, to find and extract easily recognizable samples quickly. The hierarchical cascading theory from Chap. 2 can be applied here as well.
- The work of Chap. 3 should be expanded in several ways. First, in the 2–4b precision range, several circuit and in-memory techniques become feasible, offering significant gains compared to conventional digital compute. Finding ways to further optimize deep learning kernels operating in this range of precision is still an active area of research. Other hardware-technologies, such as RRAM-cells or in-memory-compute Flash, might benefit from moving towards other network topologies, such as depthwise-separable kernels. Second, also in conventional computing, the field should move towards more hardware–software co-optimizations. Today still, network designers focus only on maximizing a network's statistical efficiency (its model size and number of operations) through automated, genetic optimization. This does not take hardware architecture, processing technology (digital, analog), or quantization into account, nor more obvious entities such as feature map sizes. A real optimal, minimum energy, model might require more operations and a larger model size than what is currently considered SotA. Hence, the field should move towards minimizing the real relevant metric: energy consumption at a given accuracy. Better hardware and energy models should be developed and combined with the automated network design platforms in order to iterate towards such minimal energy systems.
- All techniques discussed in this book could also be applied, maybe in an adapted form, to DNNs and LSTMs.
- The chips designed in this book are limited to deploying neural networks, once they have been trained offline on a HPC-system. Further research should go towards methods for reduced-precision on-chip training. A functionality that could be useful in edge devices to personalize devices to their owner or to calibrate sensors.
- The work of Chaps. 4 and 5 can be taken further by comparing the proposed DVAFS technique to the bit-serial resolution scalable techniques of Lee et al. (2018).
- The work of Chap. 6 can be improved in several ways, several ideas for which are listed at the end of the chapter. Apart from architectural optimizations, the architecture can be reused in designs with SRAM or RRAM- or Flash-based crossbars. Another examples of future work would be to expand the architecture into an XNOR-net compatible format or into a multi-bit implementation.

References

Ando K, Ueyoshi K, Orimo K, Yonekawa H, Sato S, Nakahara H, Ikebe M, Asai T, Takamaeda-Yamazaki S, Kuroda T, et al (2017) Brein memory: a 13-layer 4.2 k neuron/0.8 m synapse binary/ternary reconfigurable in-memory deep neural network accelerator in 65 nm CMOS. In: Symposium on VLSI circuits, 2017. IEEE, pp C24–C25

Andri R, Cavigelli L, Rossi D, Benini L (2016) Yodann: an ultra-low power convolutional neural network accelerator based on binary weights. In: IEEE computer society annual symposium on VLSI (ISVLSI), 2016. IEEE, pp 236–241

Bong K, Choi S, Kim C, Kang S, Kim Y, Yoo HJ (2017) 14.6 a 0.62 mw ultra-low-power convolutional-neural-network face-recognition processor and a CIS integrated with always-on haar-like face detector. In: IEEE International Solid-State Circuits Conference (ISSCC), 2017. IEEE, pp 248–249

Esser S, Merolla P, Arthur J, Cassidy A, et al (2016) Convolutional networks for fast, energy-efficient neuromorphic computing. In: Proceedings of the national academy of sciences

Han S, Mao H, Dally WJ (2016) Deep compression: compressing deep neural network with pruning, trained quantization and Huffman coding. In: International Conference on Learning Representations (ICLR)

Huang G, Che D, Li T, Wu F, van der Maaten L, Weinberger K (2017) Multi-scale dense networks for resource efficient image classification. arXiv preprint arXiv:170309844, submitted to ICLR 2018

Lee J, Kim C, Kang S, Shin D, Kim S, Yoo HY (2018) Unpu: a 50.6 tops/w unified deep neural network accelerator with 1b-to-16b fully-variable weight bit-precision. In: International Solid-State Circuits Conference (ISSCC)

Moons B, Verhelst M (2016) A 0.3-2.6 tops/w precision-scalable processor for real-time large-scale convnets. In: Proceedings of the IEEE symposium on VLSI circuits, pp 178–179

Moons B, Uytterhoeven R, Dehaene W, Verhelst M (2017) Envision: a 0.26-to-10 tops/w subword-parallel dynamic-voltage-accuracy-frequency-scalable convolutional neural network processor in 28nm FDSOI. In: International Solid-State Circuits Conference (ISSCC)

Index

© Springer Nature Switzerland AG 2019
B. Moons et al., *Embedded Deep Learning*,
https://doi.org/10.1007/978-3-319-99223-5

Printed in the United States
By Bookmasters